The Great Barrier Reef

The Great Barrier Reef is located along the coast of Queensland in north-east Australia and is the world's largest coral reef ecosystem. Designated a World Heritage Area, it has been subject to increasing pressures from tourism, fishing, pollution and climate change, and is now protected as a marine park. This book provides an original account of the environmental history of the Great Barrier Reef, based on extensive archival and oral history research.

It documents and explains the main human impacts on the Great Barrier Reef since European settlement in the region, focusing particularly on the century from 1860 to 1960 which has not previously been fully documented, yet which was a period of unprecedented exploitation of the ecosystem and its resources. The book describes the main changes in coral reefs, islands and marine wildlife that resulted from those impacts.

In more recent decades, human impacts on the Great Barrier Reef have spread, accelerated and intensified, with implications for current management and conservation practices. There is now better scientific understanding of the threats faced by the ecosystem. Yet these modern challenges occur against a background of historical levels of exploitation that is little-known, and that has reduced the ecosystem's resilience. The author provides a compelling narrative of how one of the world's most iconic and vulnerable ecosystems has been exploited and degraded, but also how some early conservation practices emerged.

Ben Daley is Lecturer in Environmental Management in the Centre for Development, Environment and Policy at the School of Oriental and African Studies (SOAS), University of London, UK. He was previously a researcher at the School of Earth and Environmental Sciences, James Cook University, Australia.

Earthscan Oceans

Governing Marine Protected Areas
Resilience through diversity
Peter J.S. Jones

Marine Policy
An introduction to governance and international law of the oceans
Mark Zacharias

The Great Barrier Reef
An environmental history
Ben Daley

Marine Biodiversity Conservation
A practical approach
Keith Hiscock

For further details please visit the series page on the Routledge website:
http://www.routledge.com/books/series/ECOCE

The Great Barrier Reef

An environmental history

Ben Daley

First published 2014
by Routledge
2 Park Square, Milton Park, Abingdon, Oxon OX14 4RN

and by Routledge
711 Third Avenue, New York, NY 10017

First issued in paperback 2017

Routledge is an imprint of the Taylor & Francis Group, an informa business

© 2014 Ben Daley

The right of Ben Daley to be identified as author of this work has been asserted by him in accordance with sections 77 and 78 of the Copyright, Designs and Patents Act 1988.

All rights reserved. No part of this book may be reprinted or reproduced or utilised in any form or by any electronic, mechanical, or other means, now known or hereafter invented, including photocopying and recording, or in any information storage or retrieval system, without permission in writing from the publishers.

Trademark notice: Product or corporate names may be trademarks or registered trademarks, and are used only for identification and explanation without intent to infringe.

British Library Cataloguing-in-Publication Data
A catalogue record for this book is available from the British Library

Library of Congress Cataloging-in-Publication Data
Daley, Ben.
The Great Barrier Reef : an environmental history / Ben Daley.
   pages cm. – (Earthscan oceans)
Includes bibliographical references and index.
   1. Great Barrier Reef (Qld.) 2. Coral reef ecology – Australia – Great Barrier Reef (Qld.) 3. Coral reefs and islands – Australia.
I. Title.
QE566.G7D35 2014
578.77′89476–dc23   2014005387

ISBN 13: 978-1-138-09570-0 (pbk)
ISBN 13: 978-0-415-82439-2 (hbk)

Typeset in Goudy
by HWA Text and Data Management, London

Contents

List of figures vii
List of tables ix
Foreword x
Acknowledgements xii
List of abbreviations xiii

1 Introduction 1

2 Reconstructing changes in the Great Barrier Reef 15

3 The natural context of changes in the Great Barrier Reef 33

4 The spread of European settlement in coastal Queensland 43

5 The *bêche-de-mer*, pearl-shell and trochus fisheries 55

6 Impacts on marine turtles 72

7 Impacts on dugongs 95

8 Impacts on whales, sharks and fish 113

9 The impacts of coral and shell collecting 128

10 The impacts of guano and rock phosphate mining 155

11 The impacts of coral mining 164

12	Other impacts on coral reefs	184
13	Changes in island biota	201
14	Conclusion	224
	References	229
	Index	246

Figures

1.1	The geographical extent of the Great Barrier Reef	2
1.2	The Great Barrier Reef World Heritage Area (GBRWHA) and the Great Barrier Reef Catchment Area (GBRCA)	3
4.1	(a) Area of sugar cane grown in Queensland, 1864–1988; (b) Fertiliser application in the Great Barrier Reef Catchment Area (GBRCA), 1900–1980	49
5.1	(a) Weights and (b) values of *bêche-de-mer* harvested in Queensland, 1880–1889	59
5.2	Values of *bêche-de-mer* harvested in Queensland, 1901–1940	60
5.3	A *bêche-de-mer* lugger near Green Island, c.1931	61
5.4	The collection of pearl-shell aboard a Queensland lugger	63
5.5	(a) Weights of pearl-shell harvested in Queensland, 1890–1940; (b) Values of pearl-shell harvested in Queensland, 1890–1940	67
6.1	Exports of tortoise-shell from Queensland, 1871–1938	74
6.2	Numbers of tortoise-shelling vessels registered in Queensland, 1895–1906	75
6.3	Exports of turtles from Queensland, 1867–1902	77
6.4	Loading a green turtle onto a boat using a winch, near Masthead Island, 1900s	79
6.5	Numbers of green turtles harvested in the Capricorn-Bunker Group, 1925–1949	81
6.6	A turtle-fishing party on the Fitzroy River, c.1930	82
6.7	Turtle-riding at Masthead Island, 1900s	84
6.8	Flatback turtles overturned on a resort island beach, 1900s	85
6.9	Turtle-riding at Heron Island, c.1930	85
7.1	Locations of individual dugong fisheries in Queensland, 1840–1970	97
7.2	Periods of operation of individual dugong fisheries in Queensland, 1840–1970	98
7.3	Quantities of dugong oil exported from Queensland, 1870–1902	99
7.4	Numbers of dugongs caught in Moreton Bay, 1884–1938	104
7.5	Dugongs caught at Burrum Heads, c.1937	105
8.1	A whale captured in east Australian waters, 1950s	115

8.2	A large stingray captured in the Great Barrier Reef, c.1930	123
9.1	Coral collecting at Masthead Island, c.1900	131
9.2	The first coral reefs and foreshores protected by legislation, in 1933	133
9.3	Assorted coral displayed at the Green Island kiosk, c.1940	138
9.4	Tourists gathering coral specimens from Heron Island reef, c.1930	139
9.5	Display of Great Barrier Reef coral for the Qantas office in Tokyo, 1961	140
9.6	Coral collecting areas in the Great Barrier Reef, 1962–1969	141
9.7	Commercial coral collectors working at Double Island reef, c.1930	143
9.8	Giant clam shells at Orpheus Island, 1967	150
10.1	Guano and rock phosphate mining locations in the Great Barrier Reef	157
11.1	Church built from burnt coral at Kobbura outstation, Fitzroy Island, c.1900	166
11.2	Coral mining locations in the Great Barrier Reef, 1900–1940	167
11.3	The location of the coral mining operation at Snapper Island	169
11.4	Letter and sketch map showing Averkoff's coral mining site at Alexandra Reef, 1929	173
11.5	Sketch map accompanying Sanders' application to mine coral from Sudbury Cay, 1930	174
11.6	Sketch map showing Garner's application to mine coral from Kings Reef No. 2 Area, 1937	176
11.7	Numbers of coral and shell-grit licences issued in Queensland between 1931 and 1968	177
11.8	The jetty at Upolu Cay used for loading material mined from the cay, c.1933	178
11.9	Alexandra Reef, near Port Douglas, 2003	179
12.1	(a) The Heron Island reef before the construction of the boat channel, 1948; (b) the same reef after the construction of the boat channel, August 1971	186
12.2	The boat channel created at Lady Musgrave Island reef, 1966	187
12.3	The access track for the lighthouse supply vessel at North Reef, 1960	188
12.4	A jetty constructed at Green Island, c.1956	193
12.5	The ten major island tourist resorts of the Great Barrier Reef, 1940	195
12.6	Construction of the Hayman Island Resort, c.1962	198
13.1	(a) Numbers of coconut palms planted in the Great Barrier Reef, 1892–1900; (b) Coconut exports from Queensland, 1905–1911	203
13.2	The distribution of island coconut palm plantations in northern Queensland, 1898	204
13.3	The coconut palm plantation at Palm Island, c.1920	206
13.4	The airstrip at Lindeman Island, October 1963	213

Tables

2.1	Some documentary sources relating to the Great Barrier Reef	17
2.2	Some visual representations relating to the Great Barrier Reef	18
2.3	Some oral sources relating to the Great Barrier Reef	19
3.1	Main types of coral reef classified by geomorphological characteristics	37
5.1	Species and values of *bêche-de-mer* harvested in Queensland, 1890	56
6.1	The IUCN Red List status of the marine turtle species of the GBRWHA	73

Foreword

The Great Barrier Reef, the largest coral reef ecosystem on earth, gained World Heritage listing in 1981 as 'the most impressive marine area in the world'. At the time of inscription, the IUCN evaluation stated '…if only one coral reef site in the world were to be chosen for the World Heritage List, the Great Barrier Reef is the site to be chosen'.

Since then, the Great Barrier Reef has been recognised as one of the best managed marine areas in the world. The 2004 Zoning Plan for the Great Barrier Reef Marine Park increased the proportion of the Marine Park that was highly protected by 'no-take' zones from less than 5 per cent to more than 33 per cent. This rezoning protected representative examples of each of the Reef's 70 bioregions (broad habitat types) occurring across its entire area of some 340,000 km^2 including the surrounding waters.

This rezoning was hailed as a new benchmark in the systematic conservation planning of marine protected areas – an example of international best practice. It won numerous national and international awards.

Nonetheless, scientific information has now established that the Great Barrier Reef has been in serious decline for decades. The causes of the decline are well known: pollution from coastal development and agricultural runoff, extreme weather events and the impacts of climate change, particularly increasing temperature and ocean acidification.

The World Heritage Committee considered the state of conservation of the Great Barrier Reef World Heritage Area in 2012, and noted with concern a critical report from the World Heritage Centre/IUCN reactive monitoring mission that visited Australia earlier that year. UNESCO is scheduled to make a decision about whether to inscribe the Great Barrier Reef on the List of World Heritage in Danger at the time of writing in 2014.

Global concerns about the future of coral reefs increasingly and appropriately focus on climate change. Nonetheless, it is important to remember that the Great Barrier Reef ecosystem abuts the eastern coast of Queensland over some 14 degrees of latitude. Land modification associated with the increased human population, urban development and agricultural expansion in the Great Barrier Reef catchment area has reduced the quality of water flowing into the Great

Barrier Reef lagoon. This area comprises some 25 per cent of the land area of Queensland, and is home to more than one million people in addition to the two million tourists that visit the region each year.

This book shows that the adverse impacts of coastal development and primary industries on the integrity of the Great Barrier Reef ecosystem have their genesis in the development of Queensland since European colonisation in the nineteenth century. Through a meticulous analysis of archival materials, official sources, publications, photographs and oral history evidence, this book documents for the first time how the colonisation of Queensland and the resultant growth of primary industries have contributed to the decline of the Great Barrier Reef. These primary industries have variously included: fisheries for bêche-de-mer, clam, coral, dugong, finfish, pearl-shell, prawns, sharks, shell-collecting, trochus, turtle and whales; mining for coral, guano and rock-phosphate; as well as agriculture, including the sugar cane and beef industries. Although most of the Great Barrier Reef islands are now protected as national parks, many islands were impacted by mining, the creation of coconut palm plantations, vegetation clearance, over-grazing by goats and the introduction of exotic vegetation.

Clearly marine park zoning alone is not enough to save the Great Barrier Reef, despite current zoning being world's best practice. This book provides important historical evidence that demonstrates that the detrimental impacts of increased human population, urban development and agricultural expansion in the Great Barrier Reef catchment area will have to be ameliorated to increase the resilience of the Great Barrier Reef World Heritage Area to the ravages of climate change.

Professor Helene Marsh
Distinguished Professor of Environmental Science
James Cook University
Townsville, Australia
June, 2014

Acknowledgements

I am very grateful for the support, advice, encouragement and assistance given by my PhD supervisors, Dr Peter Griggs and Professor Helene Marsh, both at the School of Earth and Environmental Sciences, James Cook University, Australia. Their insights and expertise in this subject continue to guide me. I am also grateful for the contribution made by Dr David Wachenfeld of the Great Barrier Reef Marine Park Authority (GBRMPA). The research was funded by an Australian Postgraduate Award (Industry) of the Australian Research Council, with additional support from GBRMPA and James Cook University. I have also received funding for this research from SOAS, University of London, where I am currently based, as well as encouragement from my colleagues in the Centre for Development, Environment and Policy. I thank these individuals and organisations for their support.

Many individuals have contributed data or technical advice to my research. I am particularly grateful to the anonymous oral history informants who participated in qualitative interviews. I also thank those individuals who suggested informants, provided materials, corrected some of my errors and misunderstandings, or otherwise contributed to the research. In particular, staff of the Australian Museum, Bowen Historical Society, Cairns Historical Society, Fryer Library, GBRMPA Library, James Cook University Library, John Oxley Library, Mitchell Library, National Library of Australia, Queensland Museum, Queensland State Archives, State Library of Queensland and University of Queensland Library provided assistance in locating documents, photographs and oral history materials.

I am grateful for the kindness and support given by the Archdeacon family, Leah Talbot, Patrick Cooke, Melissa Nursey-Bray, Debra Stoter, Robert Rutten, Tania Cobham, Andrew Bryant, Kate Hannon, Greta Galloway, Alan Webster, Christine Ritchie, Lesley Newman, David Gillieson and Elaine Harding.

I am very grateful to Tim Hardwick and Ashley Wright at Routledge and Holly Knapp at HWA for their advice, encouragement and patience during the process of writing this book.

This book is dedicated to my family, with gratitude.

Abbreviations, acronyms and measurements

Abbreviations and acronyms

ACRS	Australian Coral Reef Society
AGPS	Australian Government Publishing Service
AIMS	Australian Institute of Marine Science
ANFB	Australian National Film Board
ANU	Australian National University
A.U.S.N.	Australasian Union Steam Navigation
CHS	Cairns Historical Society
COTS	Crown of Thorns Starfish
CRC	Co-operative Research Centre
CSIRO	Commonwealth Scientific and Industrial Research Organisation
FAO	Food and Agriculture Organisation
GBR	Great Barrier Reef
GBRC	Great Barrier Reef Committee
GBRCA	Great Barrier Reef Catchment Area
GBRMP	Great Barrier Reef Marine Park
GBRMPA	Great Barrier Reef Marine Park Authority
GBRWHA	Great Barrier Reef World Heritage Area
IIB	Island Industries Board
IUCN	International Union for Conservation of Nature
JCU	James Cook University
JOL	John Oxley Library
MCMC	Mossman Central Mill Company
NADC	Northern Australia Development Committee
NLA	National Library of Australia
NPA	National Parks Association of Queensland
NQNC	North Queensland Naturalists' Club
OHC	Oral History Cassette
QBSES	Queensland Bureau of Sugar Experimental Stations
QDAIA	Queensland Department of Aboriginal and Islander Affairs
QDAS	Queensland Department of Agriculture and Stock

QDHM	Queensland Department of Harbours and Marine
QDNA	Queensland Department of Native Affairs
QDPI	Queensland Department of Primary Industries
QEPA	Queensland Environmental Protection Agency
QGITB	Queensland Government Intelligence and Tourist Bureau
QGTB	Queensland Government Tourist Bureau
QNPWS	Queensland National Parks and Wildlife Service
QPD	Queensland Parliamentary Debates
QPP	Queensland Parliamentary Papers
QPWS	Queensland Parks and Wildlife Service
QSA	Queensland State Archives
QSPCA	Queensland Society for the Prevention of Cruelty to Animals
QVP	Queensland Votes and Proceedings
SCQ	Statistics of the Colony of Queensland
SLNSW	State Library of New South Wales
SLQ	State Library of Queensland
SPCK	Society for the Promotion of Christian Knowledge
SSQ	Statistics of the State of Queensland
SST	Sea surface temperature
UNEP	United Nations Environment Program
UNESCO	United Nations Educational, Scientific and Cultural Organization
UQ	University of Queensland
WTWHA	Wet Tropics World Heritage Area

Measurements

Where used in the text, measurements have been cited as they appear in the original source. The following conversion factors may be used:

Weight and volume

1 ton = 20 cwt (approximately 1,016 kg)
1 cwt (hundredweight) = 112 lb (approximately 51 kg)
1 qtr (quarter) = 28 lb (approximately 13 kg)
1 lb (imperial pound) = 16 oz (approximately 0.5 kg)
1 oz (ounce) = approximately 30 g
1 gal (imperial gallon) = approximately 4.5 litres

Length, depth and area

1 ft (foot) = 12 inches (approximately 0.3 m)
1 in (inch) = approximately 2.5 cm
1 fathom = approximately 1.8 m
1 hectare = 10,000 square metres

Monetary values

Monetary values are stated as they appear in the original sources and no conversion factors have been applied. The following symbols denote imperial currency values:

£1 (imperial pound) = 20 s
1 s (imperial shilling) = 12 d (imperial pence)
1 d (imperial penny)

Those units were used in Australia until 14 February 1966, when the imperial currency was replaced with the modern, decimal system of dollars and cents. In this book, the modern symbol ($) refers to the Australian dollar. In 1966, following decimalisation, a conversion for the Australian dollar (in relation to the imperial pound) was £1 = approximately $2.50.

Chapter 1

Introduction

The Great Barrier Reef

The Great Barrier Reef is the largest complex of coral reefs and associated habitats on Earth (Figure 1.1). The ecosystem extends for over 2,200 kilometres along the north-eastern coast of Australia, containing around 2,900 coral reefs and representing one of the most biologically diverse ecosystems known to exist (Hutchings et al., 2009; GBRMPA, 2013). Although there has been some debate about the age of the ecosystem and of its more ancient foundations, the modern Great Barrier Reef is a young structure in geological terms, having formed during the last 10,000 years of the Holocene epoch (Hopley, 2009; Hopley et al., 2007). Consequently, its modern reefs have always existed in relation to humans, supporting the subsistence economies of coastal Indigenous Australians and containing many places of cultural and spiritual significance. After European settlement commenced in Australia, the ecosystem played an important role in the colonial development of Queensland and its resources were subjected to more intensive exploitation (Bowen, 1994; Bowen and Bowen, 2002).

Although it lies in Australian waters, the significance of the Great Barrier Reef extends beyond Australia. The coral reefs and associated habitats of the region were first protected by the creation of the Great Barrier Reef Marine Park (GBRMP) in 1975. Subsequently, in 1981, the United Nations Educational, Scientific and Cultural Organization (UNESCO) acknowledged the outstanding universal value and global significance of the ecosystem by creating the Great Barrier Reef World Heritage Area (GBRWHA) (Lucas et al., 1997). The extent of the GBRWHA is approximately 348,000 km^2, forming one of the largest and best known World Heritage Areas in the world (Figure 1.2). The Great Barrier Reef Marine Park Authority (GBRMPA), the lead agency responsible for the management and conservation of the GBRWHA, has faced considerable challenges in managing multiple human activities across the vast area of the ecosystem, and the Great Barrier Reef is regarded as one of the best managed coral reef ecosystems in the world (Wachenfeld et al., 1997; Lawrence et al., 2002).

2   Introduction

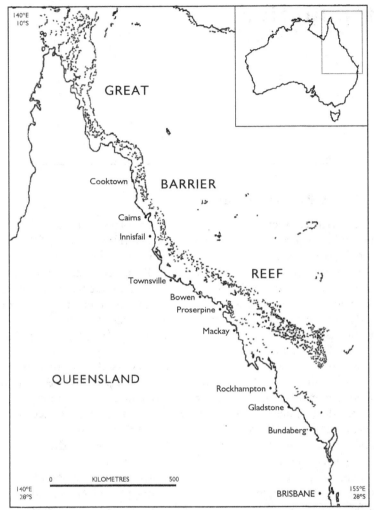

Figure 1.1 The geographical extent of the Great Barrier Reef. Source: Author, adapted from Wachenfeld et al. (1997, p4)

The Great Barrier Reef is a large, complex, dynamic entity that is difficult to define precisely. It has been defined in various ways since Matthew Flinders first used the term 'Great Barrier Reef' in 1802 (Bowen and Bowen, 2002); Maxwell (1968) listed several of those definitions in his *Atlas of the Great Barrier Reef*. The term 'Great Barrier Reef Province' refers to the coral reefs of eastern Australia and Torres Strait, one of seven coral reef provinces in the south-western Pacific Ocean. The 'Great Barrier Reef Region' describes the large coral reef area that was initially designated for protection under Australian law; in 1975, that term was replaced by the GBRMP, which extends northwards as far as the latitude of

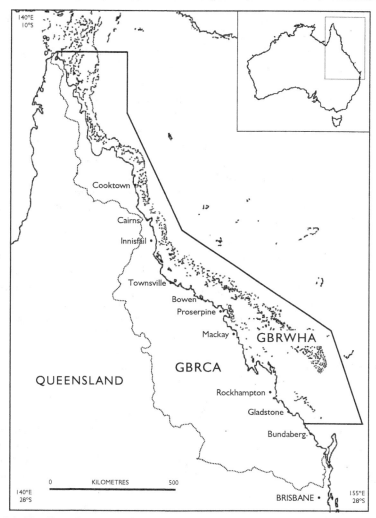

Figure 1.2 The Great Barrier Reef World Heritage Area (GBRWHA) and the Great Barrier Reef Catchment Area (GBRCA). Source: Author, adapted from Wachenfeld et al. (1997, p4); Furnas (2003, p2); GBRMPA (2009)

Cape York but excludes the coral reefs of Torres Strait. The GBRWHA occupies approximately the same area as the GBRMP, although some variations exist in their coastal boundaries: the GBRWHA also includes the islands of the Great Barrier Reef, while the GBRMP consists of the marine environment alone (Lucas et al., 1997, pp36–7, 99). All of these terms are found in the scientific literature of the Great Barrier Reef, reflecting the problem of adequately describing the boundaries of this ecosystem. Yet most historical sources predate the definitions of the GBRMP and the GBRWHA, and many simply use the general term 'Great

Barrier Reef'. In this book, a broad definition of the Great Barrier Reef is used, one that encompasses the GBRWHA (including its islands) and the adjacent areas of Torres Strait, Hervey Bay and Moreton Bay, since some important historical changes occurred in each of those parts of the ecosystem.

Conceptions of the Great Barrier Reef held by Indigenous Australians can differ significantly from those of non-Indigenous Australians. Many Indigenous Australians regard the Great Barrier Reef as part of traditional 'sea country'; some also regard it as a sacred place whose importance is reflected in creation stories. Some marine animals found in the Great Barrier Reef, including dugongs and marine turtles, have formed a vital part of the social practices and cultural identities of some coastal Indigenous Australian communities. Given the significance of this ecosystem for biodiversity conservation, human use of the GBRWHA now raises important questions about self-determination, participation and co-management of coastal and marine resources by Indigenous Australians. Recently, scholars using postcolonial approaches (amongst others) have produced new interpretations of colonisation in Australia, developed better narratives of contact and resistance, and highlighted the need for more inclusive accounts of environmental history, particularly for settler societies. Their insights suggest that accounts of the environmental history of the Great Barrier Reef that exclude Indigenous Australian perspectives are, at best, partial and incomplete (Loos, 1982; Smyth, 1994; Jacobs, 1996; Reynolds, 2003).

However the Great Barrier Reef is defined, the ecosystem does not exist in isolation: it is closely interconnected with its adjacent environments. In particular, the GBRWHA is strongly influenced by its sources of freshwater, sediments and nutrients, especially the 35 drainage basins of eastern Queensland that form the Great Barrier Reef Catchment Area (GBRCA) (Figure 1.2). The GBRCA includes around 25 per cent of the land area of Queensland, and runoff from that area represents a major input to the GBRWHA. Therefore, the GBRWHA and the GBRCA form an interconnected unit: environmental changes in the GBRWHA may be closely related to both human activities and environmental changes in the GBRCA (Furnas, 2003). Significant land uses in the GBRCA include rangeland cattle grazing, forestry, sugar cane farming, cultivation of bananas and other tropical fruits, aquaculture and mining. In addition, rapid urban development has occurred in parts of coastal Queensland, and over one million people now live in the GBRCA. Economic development in coastal Queensland has also been significant, including the growth of the commercial fishing, shipping and tourism industries in the GBRWHA, with tourism now attracting around two million visitors per year. The cumulative effects of all of those activities – in both the GBRCA and the GBRWHA – have prompted concerns about the extent to which they have contributed to the large-scale degradation of the Great Barrier Reef, particularly its nearshore habitats.

The decline of the Great Barrier Reef

Concerns about the condition of the Great Barrier Reef (and of other coral reefs worldwide) increasingly focus on the global-scale threats to these ecosystems presented by climate change, especially the effects of coral bleaching, ocean acidification and sea-level rise. Such concerns are acute since it is now recognised that, under conditions expected to occur in the twenty-first century, global warming and ocean acidification will compromise the ability of corals to form robust carbonate skeletons, with the result that coral reef ecosystems are expected to become less diverse and to have reduced capacity to maintain carbonate reef structures (Hoegh-Guldberg et al., 2007, p1737). In addition, those broad-scale changes are likely to interact synergistically with other impacts, exacerbating the effects of water pollution and disease, and thereby forcing coral reefs closer towards thresholds for functional collapse. Consequently, rapid climate change in conjunction with a range of other impacts is expected to lead to the widespread degradation and destruction of corals and coral reefs worldwide, with serious implications for tourism, reef-associated fisheries, coastal protection and human communities dependent upon reef resources (Hoegh-Guldberg, 1999; Lough, 1999; Hoegh-Guldberg et al., 2007; Veron, 2009). Within that broader context, the resilience of coral reef ecosystems to the effects of climate change could be enhanced if other (regional and local) environmental stresses are minimised. From this perspective, it becomes important to characterise and understand the extent to which the Great Barrier Reef is – and has already been – affected by those other environmental stresses, including historical ones.

Although the current, global-scale threats to coral reef ecosystems – especially those due to climate change – are formidable ones, the deterioration of coral reefs may have commenced much earlier than those threats were first recognised. Many reports suggest that the condition of the Great Barrier Reef has declined since European settlement commenced in Queensland, as a result of direct exploitation and the development of adjacent coastal land. In particular, the terrestrial runoff of sediments, nutrients and other pollutants has probably caused a substantial deterioration of water quality in parts of the Great Barrier Reef, and those effects – together with the over-exploitation of reef resources – have significantly degraded some nearshore coral reefs and seagrass communities. The Queensland Environmental Protection Agency (QEPA, 1999, p5.4) reported that nutrient inputs to the Great Barrier Reef lagoon have increased substantially over the decadal timescale due to extensive land clearing, catchment development and coastal runoff, and that the relatively enclosed and shallow nature of the lagoon makes it relatively susceptible to the effects of eutrophication and deteriorating water quality. Williams (2001, pp3–4) argued that sediment discharges have increased by three or four times, nitrogen discharges have doubled and phosphorus quantities have increased by six to ten times since 1800; as a result, the impacts of terrestrial runoff of nutrients and sediments on coastal parts of the GBRWHA – due to both past and current land use practices – have become a significant cause for concern. The impacts of coastal runoff are most significant for nearshore reefs and seagrass beds within 20 kilometres of the coast, with the most severely impacted areas lying between Port Douglas and Hinchinbrook Island, and between Bowen and Mackay (Williams et al., 2002, p1).

The Commonwealth of Australia Productivity Commission (2003, pp xxviii, 37, 42) has acknowledged evidence of an increase in sediment and nutrients entering the Great Barrier Reef lagoon since European settlement, due to the runoff of sediments, nutrients and chemicals from agricultural and pastoral land, especially as a result of cattle grazing and crop production. This has led to the decline of corals, seagrass communities and fish populations. The Great Barrier Reef Protection Interdepartmental Committee Science Panel (2003, pp2, 9, 12–13) also found evidence of accelerated erosion and a large increase in the delivery of nutrients to the Great Barrier Reef over pre-1850 levels, with consequent disturbance of the ecological function of inshore coral reefs. The report stated that some areas of the Great Barrier Reef – those most affected by terrestrial runoff – now appear to be degraded and/or slow to recover from natural disturbances such as tropical cyclones. In addition to the effects of deteriorating water quality, the degradation of the Great Barrier Reef has also occurred due to the over-exploitation of reef organisms, leading to the depletion of resources at Langford, Heron, North West, Tryon and Lady Musgrave reefs, and near Dingo, Four Mile and Kurrimine Beaches (QEPA, 1999, p5.13). Particular damage has been caused by commercial and recreational shell collecting, commercial coral collecting, aquarium fish collecting and *bêche-de-mer* (*trepang* or sea cucumber) collecting (QEPA, 1999, p5.27).

Besides these scientific and official reports, many anecdotal reports of a decline in the Great Barrier Reef have been made, attributing the degradation of coral reefs and other parts of the ecosystem to a multitude of human impacts: shipping, dredging, coastal and marine pollution, sediment and nutrient runoff, habitat destruction, coastal development, fishing, tourism and the collection of marine specimens (Lucas et al., 1997, pp65–6). The degradation of the Great Barrier Reef is considered by some observers to have occurred – or to have accelerated – in living memory. The most severe degradation is thought to have affected the nearshore habitats in the most accessible parts of the ecosystem: in the Cairns, Townsville and Whitsunday regions, which have experienced intensive human use and substantial terrestrial runoff. Given the immense ecological, economic and social importance of the Great Barrier Reef, there has been considerable scientific and public interest in either confirming or refuting those anecdotal reports of decline in the ecosystem. Furthermore, establishing the extent to which the Great Barrier Reef has changed since European settlement is important to inform the effective management of the ecosystem. However, extensive, systematic, scientific monitoring of the Great Barrier Reef commenced only around 1970, and scarce scientific data exist for the earlier period. Consequently, anecdotal claims that the ecosystem has deteriorated – especially prior to 1970 – are difficult to assess using existing scientific baselines.

Aims and approaches

In an attempt to evaluate anecdotal reports of a decline in the condition of the Great Barrier Reef for the period before extensive scientific monitoring began – especially prior to 1970 – my research used an array of qualitative methods and sources to reconstruct the environmental history of the Great Barrier Reef. Specifically, my research documents the main changes that have occurred in the coral reefs, islands and marine wildlife of the ecosystem. Many qualitative sources, including documentary and oral history materials, provide indications of the condition of the Great Barrier Reef at specific locations and at various times in the past. Archival and oral history sources, in particular, have been little used to investigate changes in the coral reefs, islands and marine wildlife of the ecosystem. A wide range of documentary materials was collected from Australian and UK archives, libraries, museums and historical societies. Those sources were used to complement and cross-reference a variety of oral history materials (both pre-existing and original): in particular, semi-structured, qualitative interviewing was used to collect new oral history evidence from informants who had observed human activities and environmental changes in the Great Barrier Reef. In addition to presenting an environmental history narrative of some of the main changes in the Great Barrier Reef since European settlement, I also evaluated the potential for qualitative methods to inform research into coastal and marine environmental history.

Coastal and marine environmental histories are not abundant in the scholarly literature, and environmental histories of coral reef ecosystems are very rare. Within the academic sub-discipline of environmental history, Australian studies comprise only a small subset, and most of those focus on the terrestrial themes of forests, soils and agriculture. There is a strong geographical bias in the literature of Australian environmental history in favour of environments in south-eastern Australia, whilst other areas have been comparatively neglected. Environmental histories of the Great Barrier Reef are very scarce: two notable works have been produced, by Bowen (1994) and by Bowen and Bowen (2002), but those focus principally on the history of exploration, environmental policy and management in relation to the Great Barrier Reef rather than on specific changes in the coral reefs and associated habitats of the ecosystem *per se*. Moreover, whilst Bowen and Bowen (2002) made extensive use of documentary materials, there is scope for a new account based on the analysis of archival sources – including some recently-available records of the former Queensland Department of Native Affairs (QDNA) – and on original oral history evidence.

Conceptual, theoretical and philosophical questions about environmental history have generated considerable debate among practitioners, and the field is characterised by a broad diversity of methods and approaches. One particularly inclusive approach is Cronon's (1992) narrative approach, which considers the task of environmental history to be, above all, the production of narratives, and which acknowledges the central role of a narrator in telling a convincing story about environmental change. Such an approach acknowledges that perceptions of environmental change reflect the diverse views of individuals in specific communities; it is therefore well suited to the collection and interpretation of qualitative materials, including the archival records of government departments and the (equally) value-laden, interpretive nature of oral history sources. In its methodology, my research was informed by the approach of Denzin and Lincoln (2000, p3), who defined qualitative research simply as 'a situated activity that locates the observer in the world'. Those authors argued that qualitative research is a distinct field of academic inquiry that is concerned with the interpretation of empirical materials in order to produce representations, such as recordings and texts; consequently, the outcome of qualitative research is itself an interpretation of reality. Like Cronon's (1992) narrative approach to environmental history, Denzin and Lincoln's (2000) approach to qualitative research emphasises the pivotal position of the researcher, whose values and attitudes fundamentally influence the research process. Therefore, those two approaches are complementary: each uses postmodern critical theory, examines the role of the narrator/researcher and emphasises the social and political contexts of representation.

Some environmental historians have emphasised the importance of reconstructing past environments in order to derive baselines that can be used to assess environmental change (Dovers, 1994; Gammage, 1994). Such reconstructions of the baseline condition of the Great Barrier Reef at the time of European settlement – if those were possible to produce – could reveal subsequent changes in the coral reefs and associated habitats of the ecosystem. Some researchers have attempted to establish this type of baseline using historical sources: Wachenfeld (1995, 1997), for example, compared historical photographs of known coral reef areas in the Great Barrier Reef with modern images, although he acknowledged that multiple methodological difficulties limit the value of that technique in reconstructing changes in coral reefs. In general, any attempt to reconstruct ecological baselines for the Great Barrier Reef is problematic, for several reasons: (a) the historical records about the ecosystem are discontinuous and extremely patchy; (b) only limited scientific monitoring of the Great Barrier Reef took place before around 1970; and (c) coral reefs are highly dynamic systems, both spatially and temporally, across a very wide range of scales. The last of these considerations has prompted some to suggest that any attempt to establish a baseline for coral reef ecosystems may be futile. Veron (2009, p36, emphasis in original), for instance, has stated that the condition of coral reefs can never be regarded as permanent, 'for there are no baselines for coral reefs, only intervals of time over which the environment *appears* not to change'. Moreover, attempts to reconstruct ecological baselines are also challenged by the postmodern view that all representations of reality are interpretive and value-laden: they do not record objective realities. Therefore, any narrative of environmental changes in the Great Barrier Reef is not definitive but represents only one of many possible readings of the historical evidence.

Although it may be impossible to reconstruct, definitively, the pre-European state of the Great Barrier Reef and the environmental changes that have occurred subsequently, it is nevertheless possible to write an environmental history narrative of changes in this ecosystem; for, as Dovers (1994, p4) has argued, there are 'good stories' to be told about the ways in which humans have interacted with their environment. The collection of evidence from documentary sources and oral histories of the Great Barrier Reef is valuable because, taken together, this evidence forms the basis of a 'convincing non-fiction' about the Great Barrier Reef that includes the perspectives of coral collectors, shell collectors, boat operators, fishers, farmers, divers, scientists, environmental managers, government officials, conservationists and naturalists, among many others (Cronon, 1992, p1373). Despite postmodern claims about the impossibility of reconstructing past environments, such an environmental history is necessary in order to provide rich, contextual descriptions of changes in the Great Barrier Reef. My account attempts to locate human activities within a context of changing environmental conditions, to allow an evaluation of human impacts to be made and to illuminate some key aspects of the changing relationship between humans, their activities and the Great Barrier Reef. In addition, in evaluating the potential for qualitative sources to inform environmental history research into an ecosystem of immense scientific interest and importance, this research has inevitably crossed disciplinary boundaries (Powell, 1996; Dovers, 2000). Interdisciplinary work of this kind can potentially offer fresh insights into the changing relationship between humans and the Great Barrier Reef since European settlement, and into the consequent environmental changes.

Scope and limitations

As an environmental history narrative, this account is concerned principally with the changing relationship between humans, their activities and the Great Barrier Reef for the historical period, which in this case is primarily the period since around 1860, when the spread of European settlement in coastal Queensland commenced. (However, some earlier European exploitation of resources in the Great Barrier Reef – such as *bêche-de-mer* fishing – had already occurred by that year.) The narrative focuses on the period until around 1960, after which an increasing body of scientific literature documenting changes in the Great Barrier Reef emerged. In general, therefore, the narrative presented here spans roughly a single century (1860–1960) of unprecedented changes in the ways in which humans encountered, experienced and exploited the Great Barrier Reef. There is some justification for focusing on this particular century of change in the ecosystem: it is a period for which scant scientific data have been collected, yet it was the century during which the most intensive exploitation of the Great Barrier Reef has ever occurred. Many historical industries operated in the region during that time – often with minimal, if any, regulation – and significant transformations of some parts of the Great Barrier Reef occurred. However, in some places, for some particular themes, and where especially informative sources were available, the narrative extends beyond that timeframe (both earlier and later). For instance, the construction of the navigation beacon on Raine Island in 1844 is included in this account, as it represents a significant (if localised) impact on one part of the Great Barrier Reef.

In general, my research focused on the area bounded by the GBRWHA, as this provided convenient limits to an extensive ecosystem that crosses administrative boundaries and that is difficult to define otherwise (Figure 1.2) However, such a definition is inevitably artificial: the GBRWHA is a recent invention. Some environmental impacts on the Great Barrier Reef derive from outside the GBRWHA, and many marine animals regularly cross its boundaries. Therefore, this research belongs within a larger geographical context. In some cases, especially where changes in marine wildlife were concerned, I used materials that related not only to the GBRWHA but that also covered the adjacent areas of Torres Strait (to the north) and Hervey and Moreton Bays (to the south). The use of such materials is justified because some large marine animals, including humpback whales, dugongs and marine turtles, range across the boundaries of the GBRWHA, and impacts have been sustained by these animals both inside and outside of those boundaries. Some early European reef fisheries, including the *bêche-de-mer* and pearl-shell fisheries, also long predated and operated across the present boundaries of the GBRWHA.

Inevitably, the scope of my research was limited by the availability of suitable materials. Nevertheless, many of the available archival materials, official reports and records of various government departments, historical books (particularly literature categorised in the 'Queensland travel and description' genre), historical photographs and oral histories (both pre-existing and original) were used extensively to obtain rich descriptions of human activities and their environmental impacts. In addition, some of the scientific reports produced as a result of the various geological, biological and ecological expeditions to the Great Barrier Reef – including the papers of the 1928–1929 Great Barrier Reef Expedition to Low Isles – and the manuscripts of notable reef scientists such as Isobel Bennett were also consulted. Interpretation of those materials relied on the use of multiple sources wherever possible, to allow evidence to be cross-referenced and to reveal errors and biases. In particular, original oral histories were compared with documentary evidence, where possible, to explore the subjectivity of informants' observations and recollections.

The scope of my research was also limited by some pragmatic considerations. Environmental changes are difficult to define, especially in a vast, complex ecosystem that is characterised by almost constant change on a wide range of geographical and temporal scales. I adopted a general, working definition of environmental change as any alteration in the physical appearance of the coral reefs, islands or their associated habitats; that broad definition was chosen as it would include the types of environmental changes most likely to have been observed by oral history informants. In contrast to major physical changes, other changes in the Great Barrier Reef – such as biological or ecological changes – were probably imperceptible to informants; nor are such changes likely to have been recorded reliably in documentary sources. Therefore, my research focuses on broad-scale, discrete, observable, physical changes in the Great Barrier Reef and neglects many other changes that may have occurred. Nonetheless, there can be little doubt that those human impacts that dominate this narrative constitute environmental changes; those activities represent the use of resources, and the modification of habitats, on an unprecedented scale in the Great Barrier Reef.

In addition to the limitations imposed by the scope of my research, several other limitations also require some explanation. My study was constrained by difficulties in obtaining suitable oral history evidence, particularly in recruiting informants who could recall visiting the Great Barrier Reef prior to the Second World War. Informants were most easily recruited in northern Queensland (particularly in the Cairns and Townsville areas), where the Great Barrier Reef lies in closer proximity to the mainland. The greater concentration of tourism activities in the Cairns, Townsville and Whitsunday areas also biased the geographical coverage of oral history evidence, for more informants encountered the Great Barrier Reef in those areas than elsewhere. Several periods of dedicated fieldwork were undertaken – including visits specifically to under-represented locations in central and southern Queensland – and telephone interviewing was also used, in an attempt to recruit informants from locations for which the oral history record was sparse.

Another limitation of my research is the uneven geographical coverage of the documentary sources describing the Great Barrier Reef. In particular, the Capricorn-Bunker Group of the Great Barrier Reef is comparatively well-documented in historical books, leaflets, films and Queensland Government reports and records, reflecting the earlier popularity of the Capricorn-Bunker Group for tourism, scientific research and naturalism in comparison with the more remote, northern parts of the Great Barrier Reef. The relative scarcity of documents describing the northern region hindered the cross-referencing of sources for this area, as there were locations for which oral histories could not be supported by documentary sources and vice versa. However, the variability in geographical coverage of documentary materials underlines the fact that European access to, and uses of, the Great Barrier Reef varied markedly along the coast of Queensland; as a result, different parts of the Great Barrier Reef inevitably have different environmental histories.

My research also faced limitations relating to the use of archival materials. McLoughlin (1999, 2000) described some of the difficulties in using archival materials in environmental research, and I encountered similar problems in my research. For example, the sequence of archival files of the Queensland Department of Harbours and Marine (QDHM) relating to coral mining in the Great Barrier Reef begins and ends abruptly, with some obvious omissions; it is likely that other, similar files were lost when the departmental offices in Brisbane were inundated during the Australia Day floods of 27 January 1974. The administration of the Great Barrier Reef prior to the formation of the GBRMP was complex, involving six different Commonwealth and Queensland

Government departments, as Bowen and Bowen (2002, p291) acknowledged, and the archival records of those various departments contain gaps and variations in their coverage. As a result of such limitations, this narrative is incomplete, and the evidence for some environmental changes is suggestive rather than conclusive. To some extent, oral history evidence could be used to fill gaps in the archival records.

A further, significant limitation of my research was the lack of an Indigenous Australian perspective towards changes in the Great Barrier Reef since European settlement, despite attempts to recruit informants from Aboriginal and Torres Strait Islander communities. Unfortunately, therefore, my environmental history does not represent Indigenous Australian perspectives towards the activities and impacts of European settlers in the Great Barrier Reef. Further research, incorporating Indigenous Australian perspectives, could provide an important, alternative perspective towards this subject. Consequently, my study contains a Eurocentric bias. I have not attempted to write a postcolonial account of Indigenous Australian resistance to European perceptions and uses of the Great Barrier Reef; nor did I attempt to investigate the ways in which Indigenous Australian interactions with the Great Barrier Reef were altered by European settlement, and with what economic, social, cultural and political consequences. Those areas of inquiry remain valuable directions for further research.

Outline of this book

The material that follows begins, in Chapter 2, with a more detailed account of the methods used in my research; the chapter outlines the sources of data available and the techniques of data collection and analysis used. It also contains a discussion of the particular issues involved in using archival materials and oral histories in my research. This is followed, in Chapter 3, by a brief overview of the natural context of changes in the Great Barrier Reef. In particular, the importance of two natural processes is emphasised: the geomorphological evolution of the continental shelf during the Holocene, and the occurrence of tropical cyclones which frequently cause damage to coral reefs. The purpose of that brief account is to demonstrate that historical changes in the Great Barrier Reef occur against a background of ongoing degradation of coral reefs in this region. Further contextual material is presented in Chapter 4, which briefly outlines the historical context for my narrative: the spread of European settlement in coastal Queensland. The chapter describes the northward expansion of European settlement and economic activities in the region, which depended on safe navigation through Queensland coastal waters, as well as on the closer settlement that accompanied the spread of sugar cane farming. The expansion of sugar cane farming was accompanied by substantial environmental degradation, as Griggs (2005, 2006, 2007, 2011) has acknowledged, due to deforestation, soil erosion and swamp drainage, which led to enhanced sediment and nutrient runoff to the Great Barrier Reef. Commercial fisheries and tourism in the Great Barrier Reef, combined with rapid coastal development, have also contributed increasingly to human impacts on the ecosystem. Therefore, Chapter 4 shows that European settlement, sugar cane farming and coastal development in Queensland have been closely interconnected with changes in the Great Barrier Reef.

Subsequent chapters present the results of my research, arranged broadly by theme. The Great Barrier Reef is a vast region in which a multitude of human activities and impacts have occurred, over varying timescales. Moreover, those activities and impacts often overlapped, both temporally and geographically, and they are sometimes difficult to classify and categorise. Consequently, it is difficult to write a chronological account of these many activities and their impacts, which have instead been grouped roughly by theme (and are not strictly chronological). One group of chapters examines different forms of exploitation of living resources in the Great Barrier Reef. Chapter 5 describes the effects of the early European reef fisheries, which exploited *bêche-de-mer*, pearl-shell and trochus. Chapter 6 examines various human impacts on marine turtles, including those due to the tortoise-shell industry, the commercial turtle fisheries, the recreational activity of turtle-riding, a turtle farming initiative and Indigenous hunting of turtles. In Chapter 7, similar impacts on dugongs are considered: those due to commercial fishing, a scheme to supply dugong products to Indigenous settlements and Indigenous hunting of dugongs. Chapter 8 covers some of the impacts sustained by humpback whales, sharks and fish due to the operation of various fisheries in the region. In Chapter 9, the effects of over-collection of corals and shells are considered, a cumulative impact on those organisms that has been previously little-documented. That group of chapters is followed by two chapters focusing on the exploitation of non-living resources: the mining of guano and rock phosphate (in Chapter 10) and of coral and coral sand (in Chapter 11). (In the case of coral mining, however, the distinction between 'living' and 'non-living' resources is not always an accurate one, since both living and non-living coral was destroyed by coral mining.) Two further chapters consider a range of other ways in which the physical and ecological habitat of coral reefs and islands was degraded or destroyed, respectively: by the clearance of access tracks and channels, military target practice operations, reef-walking and infrastructure development (in Chapter 12), and through the modification of island vegetation and fauna (in Chapter 13). Finally, the book concludes with a brief consideration of the overall significance of these multiple, unprecedented impacts on the Great Barrier Reef (Chapter 14).

Taken together, this account provides evidence of sustained, extensive damage to some coral reefs, islands and organisms of the Great Barrier Reef as a result of the over-exploitation of resources and the degradation and destruction of habitats. There is unequivocal evidence that some parts of the Great Barrier Reef have experienced severe impacts – that have varied in their location and intensity – since European settlement. Consequently, although the Great Barrier Reef remains one of the best-protected coral reef ecosystems in the world, some of its habitats were far from pristine at the time of the formation of the GBRMP in 1975. At least partially as a result of human activities, some coral reef areas have been degraded to the extent that recovery to their former condition is now unlikely, and many other reef areas have been affected to a lesser extent. Similarly, there is abundant evidence that some islands of the Great Barrier Reef – particularly Raine, Green, Magnetic, Holbourne, North West, Heron, Fairfax, Lady Musgrave and Lady Elliot Islands – were subjected to considerable exploitation and had been significantly modified by the time of the formation of the GBRMP. In addition, some of the impacts sustained by marine wildlife species in the Great Barrier Reef have been severe and, whilst their ecological consequences are not always easy to establish, it is clear that the over-exploitation of some populations and species – sometimes over the decadal timescale – has substantially increased their mortality, morbidity and vulnerability to other environmental changes. Overall, the evidence presented in this book suggests that the Great Barrier Reef was exploited earlier, for a longer period, in more locations and more intensively than has previously been documented. That story – of the multiple and unprecedented uses of, and changes in, the Great Barrier Reef since European settlement – forms the main subject of my narrative. Above all, my account illustrates the damage that can be inflicted on coral reefs and their associated habitats and species in the absence of effective environmental management.

Chapter 2

Reconstructing changes in the Great Barrier Reef

Introduction

Few scientific studies of the Great Barrier Reef cover the period 1860–1960, the one covered in this book. In contrast, qualitative sources can provide information about environmental changes in the Great Barrier Reef for that period; useful sources include both documentary (especially archival) and oral history sources, which have been little used in environmental histories of the Great Barrier Reef. My study used an array of qualitative methods and sources to reconstruct changes in the coral reefs, islands and marine wildlife of the Great Barrier Reef for that period. However, the use of qualitative methods and sources requires a systematic approach, and raises specific methodological and philosophical issues. This chapter provides an outline and explanation of the sources and methods used in my research. Two main types of qualitative source were used – documentary and oral sources – including historical books, maps, photographs, official records and reports of Queensland Government departments, newspapers and oral history interviews. However, among those sources, I emphasise the particular importance of archival and oral history evidence in my research. Some of the limitations of qualitative sources – including their subjectivity, bias and partial coverage – are also considered below, and some responses to those limitations, including cross-referencing and the use of expert, scientific advice, are described. This chapter also contains a discussion of the particular issues associated with using oral histories.

Sources of data

Two main types of qualitative source were used in my study: documentary and oral sources. However, as Ganter (1994, p12) acknowledged, 'the distinction between written and oral sources is, in any but the most pragmatic sense, an artificial one'. She acknowledged that written sources may contain as many inaccuracies and subjectivities as oral sources, and she argued that all sources must be treated with equal caution. Ganter (1994) also suggested that some qualitative sources fall into an ambiguous category – between written and oral sources – such as the transcribed evidence of Royal Commission inquiries. Therefore, she argued, no single source of data should be privileged above any other. Despite Ganter's observations, I made a distinction between documentary and oral sources in my research, because one intended outcome of the study was an evaluation of the value of different qualitative sources and methods in environmental history research for a coastal and marine environment.

Documentary sources include European and Australian written records describing various aspects of the Great Barrier Reef for the last 150 years, approximately. The most diverse and abundant of those documents are the historical books held in the collections of major Australian libraries and GBRMPA. Historical books include both fictional and non-fiction accounts of the Great Barrier Reef. Many historical leaflets – including tourist literature – also describe the Great Barrier Reef. Another significant documentary source is the official Queensland Government reports contained in the *Queensland Parliamentary Papers* (QPP), *Queensland Parliamentary Debates* (QPD) and *Queensland Votes and Proceedings* (QVP); those documents are held in the major libraries of Queensland. Export statistics for the early European reef fisheries were obtained from the *Statistics of the Colony of Queensland* (SCQ) and the *Statistics of the State of Queensland* (SSQ). A further documentary source is the records of various Queensland Government departments; those are held in Queensland Government departmental offices or in the Queensland State Archives (QSA) in Brisbane. Other documentary sources include national, regional and local newspapers, which contain journalistic accounts, and various types of manuscripts (such as the records compiled by individual scientists and naturalists). Some of the written documentary sources I used are listed in Table 2.1.

Documentary sources also include visual representations such as historical photographs, films, video-recordings, maps, posters and sketches. Such materials can contain evocative images of past conditions of the Great Barrier Reef, although the significant difficulties involved in their interpretation have been discussed by Ball and Smith (1992) and by Wachenfeld (1995). Visual representations are held in the collections of major libraries in Australia, particularly those with significant historical collections, such as the Mitchell Library in Sydney. Some materials are held in specialist collections of historical photographs, maps, films and video-recordings, as, for example, at the GBRMPA Library in Townsville, where the Historical Photographs Project forms a special photographic collection (Wachenfeld, 1995, 1997). Many methodological issues arise in the use of visual representations – especially in the interpretation of historical photographs – as discussed below. Nonetheless, some images provided valuable information about historical human activities in the Great Barrier Reef, including photographs of coral collecting. In some cases, visual representations are the only surviving sources of data for particular locations or periods. A list of the visual representations used in my research is found in Table 2.2.

Oral sources comprise spoken materials, often preserved in recorded interviews and transcripts. They include a range of types of speech: from

Table 2.1 Some documentary sources relating to the Great Barrier Reef

Source	Types of material	Locations
Historical books	Scientific texts Description and travel literature Fiction	National Library of Australia, Canberra Australian Museum Library, Sydney Mitchell Library, Sydney Queensland Museum Library, Brisbane John Oxley Library, Brisbane Fryer Library, St Lucia, Brisbane GBRMPA Library, Townsville Townsville City Library, Townsville James Cook University Library, Cairns Cairns City Library, Cairns
Historical leaflets	Tourist leaflets Tourist brochures Information pamphlets	National Library of Australia, Canberra Mitchell Library, Sydney John Oxley Library, Brisbane GBRMPA Library, Townsville
Government reports	Queensland Parliamentary Papers (QPP) Queensland Parliamentary Debates (QPD) Queensland Votes and Proceedings (QVP) Statistics of the Colony of Queensland (SCQ) Statistics of the State of Queensland (SSQ)	University of Queensland Library, St Lucia, Brisbane Cairns Historical Society, Cairns
Government records	Records of the Queensland Department of Harbours and Marine (QDHM) Records of the Queensland Environmental Protection Agency (QEPA) Records of the Queensland Department of Aboriginal and Island Affairs (QDAIA)	Queensland State Archives, Brisbane Queensland Environmental Protection Agency, Brisbane Queensland Department of National Parks and Wildlife, Cairns
Newspapers	*The Cairns Post*	*The Cairns Post* Archives, Cairns James Cook University Library, Cairns Cairns City Library, Cairns
Manuscripts	Field diaries Field notes Reports Presentation notes Book manuscripts Business records Miscellaneous correspondence	Manuscript collection, National Library of Australia, Canberra

Source: Author

Table 2.2 Some visual representations relating to the Great Barrier Reef

Source	Types of material	Location
Historical photographs	Scenic photographs Scientific photographs Aerial photographs Postcards	IMAGES1 photographic collection, National Library of Australia, Canberra Geoscape scanned aerial photographs, National Library of Australia, Canberra Australian seashores colour transparency collection, Isobel Bennett, National Library of Australia, Canberra Australian Museum Archives, Sydney Mitchell Library, Sydney John Oxley Library, Brisbane Historical Photographs Project, GBRMPA, Townsville Cairns Historical Society, Cairns
Historical films	Scenic films Scientific films	ScreenSound Australia (National Screen and Sound Archive), Canberra State Library of New South Wales, Sydney Mitchell Library, Sydney John Oxley Library, Brisbane State Library of Queensland, Brisbane
Historical video-recordings	Scenic films Scientific films	ScreenSound Australia (National Screen and Sound Archive), Canberra State Library of New South Wales, Sydney State Library of Queensland, Brisbane
Historical maps	Hydrographic survey charts Tourist maps Scientific maps	Historical Maps Collection, National Library of Australia, Canberra Mitchell Library, Sydney John Oxley Library, Brisbane Maps Unit, State Library of Queensland, Brisbane Queensland State Archives, Brisbane Cairns Historical Society, Cairns
Historical posters	Information posters Tourist posters	State Library of New South Wales, Sydney State Library of Queensland, Brisbane
Sketches	Field sketches Sketch maps *Cape Moreton* sketch maps and sketches	Manuscript collection, National Library of Australia, Canberra Queensland State Archives, Brisbane Queensland Museum Library, Brisbane

formal, structured oral history interviews to informally collected conversations, unstructured interviews, anecdotal reports and comments (Gillham, 2000). Robertson (2000) acknowledged that, while oral sources are sometimes regarded as the most unreliable and subjective of all qualitative sources, they also offer distinctive insights into the attitudes and motives of human actors. Similarly, Allen and Montell (1981, p3) stated that 'orally communicated history is a

Table 2.3 Some oral sources relating to the Great Barrier Reef

Source	Types of material	Location
Pre-existing oral histories	Audio cassettes Interview transcripts Interview recording notes	Oral history collection, National Library of Australia, Canberra Oral history collection, School of History, James Cook University, Townsville
Original oral histories	Audio cassettes Interview transcripts Interview recording notes Interview schedules	Collected at various locations in Australia, and by telephone
Anecdotal comments	Memos Field notes	Collected at various locations in Australia, and by telephone

Source: Author

valid and valuable source of historical information, as oral tradition and formal history complement one another'. Ganter (1994, pp11–12), likewise, argued that written, visual and oral sources may be equally useful – and equally problematic. Oral sources offer unique opportunities to investigate human perceptions, beliefs, feelings, intentions and memories: they can be used to explore those experiences in depth and detail. Oral evidence comprises both specially-collected, original interviews with key informants and pre-existing materials that are available for re-interrogation. Both original and pre-existing oral histories were used in my research: those sources are summarised in Table 2.3.

Major sources of data relating to the Great Barrier Reef are the collections of the following Australian institutions: GBRMPA (Townsville), the Australian Institute of Marine Science (AIMS) (Townsville), the National Library of Australia (NLA) (Canberra), ScreenSound Australia (the National Screen and Sound Archive) (Canberra) and the Australian Museum (Sydney). Those institutions contain much of the scientific literature of the Great Barrier Reef and also certain specialist oral history, photographic and film collections. The major state libraries containing sources relevant to the Great Barrier Reef include the State Library of Queensland (SLQ) (Brisbane) and the State Library of New South Wales (SLNSW) (Sydney). They hold collections of maps and films about the Great Barrier Reef. Historical libraries include the John Oxley Library (JOL) (Brisbane), the Mitchell Library (Sydney) and the Fryer Library of the University of Queensland (Brisbane). Those libraries hold specialist historical collections including many books, leaflets and photographs. Other relevant libraries include the University of Queensland Library (Brisbane), the James Cook University (JCU) Library (Cairns and Townsville), the Cairns City Library and the Townsville City Library.

Other than in libraries, qualitative sources are held in the offices of the QEPA (Brisbane) and in the QSA. The latter holds Queensland Government records, catalogued by provenance, after files have been transferred from departmental offices. Other, active files are held in Queensland Government departmental offices at several locations in Queensland. The management plans for the island National Parks of the Great Barrier Reef, for example, were obtained from the office of the Queensland Parks and Wildlife Service (QPWS) in Cairns. Access to some of those records is restricted for varying time periods, and the use of some files required special permission. Newspaper collections are held in various public libraries and archives of Queensland; I searched archives and microfilm copies of *The Cairns Post* at the newspaper's offices in Cairns, for instance. Finally, many qualitative sources – including personal photographs – are held privately and are widely distributed. Oral history informants who could recall with accuracy the period before 1980 were scarce and widely dispersed. Therefore, the availability of oral history sources for this period is limited; particular strategies were required for the collection of those multiple, diverse sources of data. My methods of data collection are described below.

Data collection

This section contains an account of the methods used to collect data from documentary and oral sources. Data were collected from many qualitative sources, since the purpose of my research was to use an array of such sources to synthesise an account of environmental changes. The value of using more than one method has been explained by Denzin and Lincoln (2000, p19):

> No single method can grasp all of the subtle variations in ongoing human experience. Consequently, qualitative researchers deploy a wide range of interconnected research methods, always seeking better ways to make more understandable the worlds of experience they have studied.

The array of qualitative methods used included some of the methods listed by Dey (1993, p2), including case study, content analysis, descriptive research, document study, field study, focus group research and oral history methods. As well as allowing the collection of richer data, the use of an array of methods provided a means of cross-referencing and correcting sources and of assessing their internal consistency.

The use of multiple methods also allowed an assessment to be made of the validity and value of different qualitative sources for reconstructing past environments; my approach sought both to reconstruct an environmental history using qualitative sources and to evaluate those sources (Hoggart et al., 2002). Some authors, such as McCracken (1988) and Brannen (1992), have argued that qualitative research should ideally use mixed methods, including both qualitative and quantitative techniques. Despite the value of such an approach, my research focused on qualitative methods, since an assessment of their potential for environmental history research was one of the intended outcomes of my study. The task of integrating qualitative findings with the extensive scientific literature of the Great Barrier Reef goes beyond the scope of my research, although there is considerable potential for further research in that area.

Collection of documentary data

Documentary data collection involved searching for and analysing historical books, leaflets, Queensland Government reports and records, newspapers and manuscripts. My search procedure used various strategies: a system of pre-defined search terms used with electronic catalogues; electronic searches of relevant call numbers to locate clusters of similar works; manual searches of index card catalogues; and consultation with professional librarians, curators and archivists. The documentary materials used, assembled together, provided a broad range of textual sources for analysis (Travers, 2001). The use of many types of documentary sources allowed for their cross-referencing and cross-validation, and these revealed differences in perceptions of environmental changes between authors. Some examples of those differences are evident in the narrative that follows. Moreover, the distinctive nature of each documentary source, and the need to use methods that adapt to the peculiarities of each source, have been acknowledged by Hakim (1987); for example, newspaper reports about the impacts of commercial dugong fishing required more cautious interpretation than the Queensland Government reports about that industry. My documentary data included a subset of data obtained from visual representations, which are described separately. The methods of data collection used for each type of document are described below.

Books

Historical books were located in the following collections: the NLA, JOL, Mitchell Library, Fryer Library, SLQ, SLNSW, JCU Library, GBRMPA Library, Cairns City Library and Townsville City Library. Several hundred historical books were identified using electronic and manual search techniques. Those works were then analysed for observations relating to the condition of, or changes in, the Great Barrier Reef. The historical books consulted were divided into four main categories: records of early European explorers, scientific texts of the Great Barrier Reef, Queensland description and travel literature, and works of fiction.

The records of many early European explorers of the Queensland coast and the Great Barrier Reef have been published in book (or microfilm) form. The records of the earliest British exploration of the Great Barrier Reef, made aboard *HMS Endeavour* in 1770, were redacted and published in the journals and maps of Captain James Cook and of Joseph Banks. Material contained in the journals of other explorers, including Bligh, King, Stokes, Flinders, Huxley, Wickham, Bunker, Owens and Stanley also describes the north-eastern coast of Australia. Histories of early European exploration of Australia (Gill, 1988) and of shipwrecks in the Great Barrier Reef (Holthouse, 1976) were also consulted.

Records of early European explorers were generally in edited form; however, copies of some original works were consulted. However, the records of some early European explorers, such as the account of the voyage of the *Astrolabe*, were not used as no English translation was available.

Scientific texts used in my research included historical books dealing with the scientific description, investigation and analysis of the Great Barrier Reef. Studies by Saville-Kent, for example – including his extensive study of the Great Barrier Reef (Saville-Kent, 1893) – represent some of the earliest scientific work devoted to the ecosystem. The scientific papers and books relating to the 1928–1929 Royal Society of London expedition to Low Isles form a significant source of data (British Museum (Natural History), 1930–1968; Yonge, 1930). Other important scientific texts included the many geological studies published and the early scientific reports of the Great Barrier Reef Committee (GBRC); those documents were searched comprehensively for evidence of changes in the Great Barrier Reef, including comparisons between studies of the same locations (Agassiz, 1898, 1913; Hedley, 1925; Hill, 1960).

An extensive body of literature categorised in the 'Queensland description and travel' genre describes the landscapes, human activities and development of Queensland during much of the colonial period. This literature also includes observations on, and recollections of, travel in the region; therefore, it offers some rich descriptions of the condition of the Great Barrier Reef written by European and Australian explorers, tourists, journalists and beachcombers. An example of this genre is *The Confessions of a Beachcomber*, by Edmund Banfield (1908), describing Dunk Island and surrounding islands. Similar books – such as *On the Barrier Reef* (Napier, 1928) and *Destination Barrier Reef* (Lock, 1955) – include descriptions of specific locations along the Queensland coast at different times. Therefore, those works were included in my data collection strategy; however, they were extremely numerous, and their titles gave little indication of their scope or content. Some more general, national-scale works about description and travel in Australia contained brief references to the Great Barrier Reef, but that category was sampled and found to reveal scarce useful information about the ecosystem.

The collection of documentary data from historical works of fiction was informed by Sharp (2000), who considered the use of literary fiction by geographers and argued that the social construction of fictional texts has been less well understood than the construction of scientific texts. She argued for more careful analysis of the content and form of fictional texts and for the distinctive voice of literary fiction to be recognised in geographical studies. Sharp (2000, p329) acknowledged the danger of misinterpreting fictional texts and stated that 'geographers are still drawing from literature those elements that reinforce the position to be argued'. She highlighted the fact that researchers may select from fictional texts only those parts which reinforce a pre-existing interpretation. In this respect, she echoed Cronon's (1992) view that environmental historians are narrators who may inevitably reach the conclusions that they expect to find.

Contextual understanding of fictional texts is crucial if those sources are to make a distinctive contribution to geographical knowledge. Sharp (2000) highlights the subjectivity inherent in analysis of fictional texts; for that reason, although fictional accounts of the Great Barrier Reef were collected, that material was used to exemplify general perceptions of coral reefs rather than analysed for historical information using the criteria that were applied to other books.

In works of historical fiction – such as *Death on the Barrier Reef* (Antill, 1952) – any descriptions of coral reefs and their associated environments and species that could be precisely located within the Great Barrier Reef were collected; however, there were very few such examples. In addition, however, pieces of descriptive text relating to unidentifiable reef locations were also collected if they represented exemplary or evocative writing about the Great Barrier Reef. Fictional texts were searched for material which illustrated changing cultural constructions, as well as changing physical environments, of the Great Barrier Reef. Fewer works of historical fiction were found than other historical books, and fewer data were taken from these works. Love (2000) has provided a useful account of the ways in which scientific accounts of the Great Barrier Reef contrast with fictional works.

Leaflets

Historical leaflets form a small but highly descriptive documentary source; those documents were primarily intended to promote the Great Barrier Reef to visitors, particularly tourists. Most of the historical leaflets consulted in my research were published by the Queensland Government Intelligence and Tourist Bureau (QGITB), which later became the Queensland Government Tourist Bureau (QGTB). Other, more scientific, leaflets were produced by the GBRC. The leaflets reflect the descriptions and cultural constructions of the Great Barrier Reef at the time of increasing tourist development in the area. The tourist leaflets required especially cautious interpretation; they contain superlative descriptions of corals and marine wildlife in an attempt to attract visitors. In contrast, the information leaflets produced by the GBRC present a less emotive image and contain more systematic coverage of the reefs.

Government reports

In my research, a distinction was made between government reports (the published, annual reports of Queensland Government departments) and government records (the unpublished correspondence and other files held in Queensland Government departmental offices and in the QSA). The reports used as data sources included the published annual reports of several Queensland Government departments, including the QDHM, the Queensland Department of Agriculture and Stock (QDAS), the QDNA and the Agent-General for Queensland in London. Materials relevant to the Great Barrier Reef were found in the annual reports of those departments, published in the *QPP* and the *QVP*. Information about the Great Barrier Reef could be contained within the reports of any department, in any year, because diverse activities such as mining, coconut palm planting, dredging and tourist infrastructure development were administered by different departments. Therefore, many departmental reports were sampled, and many reports were searched systematically and comprehensively, for references to the Great Barrier Reef.

The collection of data from government records, therefore, represented both a longitudinal study of the activities of individual departments, from year to year, and a cross-sectional collection of materials across departments for the same year. This method allowed data to be found in the years that responsibility for their collection was transferred from one department to another. In addition to the annual reports, reports of governmental debates and proceedings (for the Queensland Government) were published in the *QPD*, and export statistics for various products – such as tortoise-shell – were obtained from the *SCQ* and the *SSQ*. The *QPD*, however, were consulted for additional material on key activities in the Great Barrier Reef, but they were not searched exhaustively. Parliamentary reporting procedures suggested that the major events involving Queensland Government departments would be reported primarily in the *QPP* (and in the *QVP*, prior to the formation of the State of Queensland, in 1901) and those documents represented the most comprehensive source of data about departmental activities. Data from all of those records, in addition, was supplemented by the unpublished government records sourced from departmental offices and the QSA (see below).

Over the period 1860–1960, the reporting procedures of Queensland Government departments changed significantly and the organisation of departments altered; some, such as the QDHM and the QDNA, ceased to exist during that period. The records available in the annual reports are neither continuous nor consistent; they were intended primarily as political, not scientific, documents. Consequently, the recording and interpretation of data found in government reports requires caution, and is sometimes unsuitable for use in reconstructing environmental change. In particular, longitudinal data series contain many disjunctions. For example, in some years the reporting of total annual catches of green turtles by weight included both shell and shell meat combined; in other years the weights of shell alone were reported. Another example is the reporting of coconut palm planting, which omitted detail for the earliest plantations. Such anomalies support McLoughlin's (1999) findings that major omissions and inconsistencies occur in official dredging records. Yet government records provide valuable insights into the activities of Queensland Government departments and a source of data that, in some cases, represents the only surviving record of particular human activities in the Great Barrier Reef.

Government records

In my research, government records comprised the majority of the archival data collected. Both public and private records were sought, as distinguished by Frankfort-Nachmais and Nachmais (1992, p305). Those authors categorised four types of public records: actuarial records, political and judicial records, government documents, and media reports; of those, actuarial records (which record demographic characteristics) and political (electoral) and judicial records were not relevant to my research. However, the records of Queensland Government departments revealed important changes in the environments of the Great Barrier Reef, and media reports also contained valuable data about changes in the reefs, so those two types of archival source were sought. Private records included diaries, autobiographies and letters; the private records of key informants, including diaries and sketches, were sought in my research and in some cases were of considerable value in documenting changes in the Great Barrier Reef.

Archival data collection was an exploratory and intuitive process. The data used were found in the Australian Museum archives and the QSA. Elder et al. (1993, p11) considered the use of archival data in research and made five main observations. First, they argued that archival data are never ideal for the research intended; the researcher is challenged to use the limited materials available as resourcefully as possible. Second, archival materials reflect the concerns of their original collectors and the social and political context of that time; they do not always fit into modern categories nor reflect modern sensibilities. Third, longitudinal archives are scarce and rarely yield continuous, consistent data; cross-sectional records are far more common. Fourth, both quantitative and qualitative data are found in archival records and may require different kinds of analysis. Finally, Elder et al. (1993) argued that any rationale for using archival data should be based on the strengths of that data, not on any attempt to ignore or overcome their weaknesses. Those observations suggested that the process of archival data collection in my research should be speculative and intuitive; the archives were searched using electronic catalogues – sampled using pre-defined search terms – and also using the expertise of professional archivists.

Other insights into the use of archival sources are provided by McLoughlin (1999), who assessed the use of Australian government dredging records in environmental history research; she acknowledged that major gaps exist in the records and that a significant amount of basic information is missing from them. McLoughlin (1999) concluded that an inadequate record of historical dredging has been preserved in archival sources and she explored the tensions existing within archival theory and practice. She also noted the distinction between different types of government archival materials: published departmental annual reports, detailed operational records and correspondence files; no complete record was found in any of those types of government material. In my research, similar difficulties arose in collecting data both from published government records and from archival government reports. In particular, many of the files expected to be preserved at the QSA – including licences for coral mining areas whose existence was suggested by the logical sequence in which the licences were issued – were not found despite exhaustive searches. In addition, files that were initially expected to be found at QSA – including the earliest management plans for the island National Parks of the GBRMP – were not found there, and a strategy of searching for government records in regional departmental offices was developed in response to this problem. Some management plans, for example, were obtained instead at the departmental offices of the QPWS, in Cairns. The cases described above suggested that some useful documents may still potentially be available in regional departmental offices, or otherwise may have been lost or destroyed.

Newspapers

Newspapers contain journalistic articles about many events in the history of Queensland; a selection of newspaper reports was searched for data about reported conditions and changes in the Great Barrier Reef. Since those reports contain an enormous amount of material, for many localities, over an extensive time period, the selection was limited to those locations, dates and topics for which other sources had already indicated that useful information might be gained. For example, information about the commercial dugong fisheries was obtained from the *Brisbane Courier-Mail* and the *Sydney Morning Herald*, and evidence of coral and guano mining in the Cairns region was obtained from reports published in *The Cairns Post*. *The Cairns Post* was also searched using an index at the JCU Library, a catalogue at the Cairns Historical Society (CHS) and *The Cairns Post* archives in Cairns. Analysis of newspaper reports revealed that popular perceptions of environmental changes – for example, the impacts of coral mining on birds at Michaelmas Cay – were highly subjective and required cross-referencing with other sources of data. Details of coral mining in the Great Barrier Reef were found, for instance, but evidence of that activity gained from newspaper reports required corroboration by other evidence.

Manuscripts

Manuscripts included collections of miscellaneous documents: papers, field notes, diaries, presentations, book manuscripts, photographs and correspondence. Several relevant collections of material were found in the manuscript collection of the NLA, including the papers of the renowned marine biologist, Isobel Bennett, and of the poet, Judith Wright. Those manuscripts were searched exhaustively, as they related explicitly to the Great Barrier Reef, and they yielded many data. For example, the papers of Isobel Bennett provided details about the creation of the boat channel at Heron Island, perceived impacts of coastal development on the condition of inshore coral reefs and the growth of tourism in the Great Barrier Reef. However, inevitably, the manuscripts of different individuals varied in their relevance to my research, and typically they were not fully catalogued or indexed.

Collection of visual representations

Visual representations, in my research, included historical photographs, films, video-recordings, maps, posters and sketches. The use of paintings did not form a significant part of this research, despite Nordstrom and Jackson's (2001) argument about the value of using paintings to explore interactions between human activity and coastal change. A preliminary survey of the collections of paintings held in the Queensland Art Gallery (Brisbane) and the Cairns Art Gallery (Cairns) did not reveal sources of relevance to the Great Barrier Reef; hence, an extensive search for relevant paintings was not pursued. Furthermore, paintings were not expected to reveal accurate, place-specific representations of particular environmental changes in the Great Barrier Reef. However, a notable exception was the work of the painter, Ray Crooke, which provided information about the condition of Magnetic Island during the Second World War. Other visual sources – particularly historical photographs – showed greater potential to reveal environmental changes, although Wachenfeld (1995) has discussed the methodological difficulties that arise in their use (see below).

Visual representations – used in conjunction with other documentary sources and with oral sources – provided a comparatively small but unique data set. A sketch map of the coral mining operation at Snapper Island, for example, represents evidence of the details of an environmental change for which no other documentary source was found. While visual data were difficult to evaluate, record and interpret, they nevertheless gave a vivid impression of human impacts in the Great Barrier Reef, including coral mining, coral collecting, shell collecting, access track and channel construction, coconut palm planting, infrastructure development on islands, commercial dugong and turtle fishing, turtle-riding and whaling. Visual representations also suggested the various ways in which cultural constructions of the reefs were produced – and reproduced – during the period of European settlement. The methods used to collect data from each type of visual source are discussed below.

Photographs

Photographs are among the earliest representations of the Great Barrier Reef; they have been used specifically to record the condition of the reefs and their associated species since the work of Saville-Kent in the 1890s. The quality and abundance of historical photographs of the reefs led GBRMPA to establish its Historical Photographs Project, which involved the collection and analysis of photographs in search of evidence of environmental changes. Indeed, Saville-Kent's photographs were explicitly intended to be used for that purpose, as Wachenfeld (1995) acknowledged. However, although the

Historical Photographs Project assessed an extensive collection of images and drew many conclusions about the apparent decline – and occasionally the apparent improvement – of particular coral reefs, Wachenfeld (1995) argued that comparing historical photographic images with re-photographed, modern locations is too problematic for firm conclusions to be drawn about changes in coral reefs. Moreover, Wachenfeld's (1995) analysis suggested that different environmental changes had occurred, in different places, and that no overall trend could be identified. Therefore, in my research, the collection of data from historical photographs did not take place specifically for the purpose of re-photographing reefs, although some photographs were collected if they provided clear evidence of human influence on the coral reefs, islands and marine wildlife of the Great Barrier Reef. For example, historical photographs document the recreational activity of turtle-riding by tourists at resort islands.

Historical photographs were consulted initially at the collection of the Historical Photographs Project, held by GBRMPA, in Townsville. Many other collections of photographs were searched at the major libraries; in particular, the historical photographs collection at the JOL was searched exhaustively, and approximately 2,000 images were surveyed. In addition to the photographs at the JOL, other historical photograph collections were searched at the Mitchell Library, the NLA and the QSA. Many of the images in those collections that were relevant to the Great Barrier Reef had already been collected elsewhere; however, the search process revealed many inaccuracies in the identification, labelling and cataloguing of images: numerous images were identified with incorrect locational information, and some photographs were identified differently in different collections. Many images lacked information about the location or date of the photograph, so were of limited use in my research. Cross-verification of photograph details was carried out as far as possible.

In addition to the public collections of historical photographs, private holdings were also obtained during the process of recording oral history interviews. Informants were asked to provide relevant photographs of the Great Barrier Reef if they were able to do so. Some informants provided extensive, annotated collections of private photographs that were used during or after the interviews. One informant was able to compare their own photographs with published images of the same locations. Another use of historical photographs in interviews required informants to identify or discuss images, which sometimes prompted informants to recall, discuss or explain further details of their evidence. Finally, while the re-photography of reefs for which historical photographs exist was not an aim of my research, the re-photographed evidence of changes at Heron Island and at Low Isles, collected by Isobel Bennett, was used (with her permission) in my research since it provided unambiguous evidence of significant changes; some original images of reefs – such as Yule Point reefs – were also collected as they provided evidence of the current condition of those reefs.

Films and video-recordings

Historical films included the works of the earliest professional film-makers to document the Great Barrier Reef using both surface and underwater photography, such as Noel Monkman. The films were located in the audio-visual collections of the SLQ, the SLNSW and at ScreenSound Australia. Those collections also included video-recordings; footage preserved in each form was consulted. Many early recordings of the Great Barrier Reef were produced by the Australian National Film Board (ANFB) and those included Monkman's underwater films of coral and associated species. Other relevant films were produced for the *Australian Diary* series, including earlier footage of tourist and industrial development in north Queensland and the resorts of the Great Barrier Reef. Those were surveyed and analysed as they documented the growth of tourist resorts, such as at Green Island, and also contained evidence of human impacts on turtle populations in the Capricorn-Bunker group. Some potentially valuable historical films were not available for study because of damage to the materials; consequently, other historical films exist that could potentially yield valuable information if they are restored.

Maps

Many historical maps of the Great Barrier Reef have been produced; they are among the earliest surviving European records of the existence and nature of the reefs. Early European exploration and navigation depended on the development of accurate hydrographic charts; some of those are preserved in the collections of many Australian libraries and archives, often in specialist map collections. Besides hydrographic charts, other maps of the Queensland coast were used, including scientific, tourist and road maps. Maps were analysed for data about particular environments in the Great Barrier Reef and they revealed the nature of previous coastal and island vegetation, types of substrate, growth of settlements, presence of discoloured seawater and locations of fringing reefs. Maps were used in conjunction with other materials, where possible, to reveal evidence of environmental changes; for example, maps of Raine Island, Low Isles and North Reef indicated the impacts of guano mining, the destruction of the '*Porites* Pond' and the blasting of an access channel through the coral reef, respectively. Yet, overall, the value of historical maps for my research was limited: many represented the reefs and coastline with varying degrees of precision, and often with inconsistent or incorrect place names, perhaps due to the difficulty of conducting accurate cartography in the challenging environment of the Great Barrier Reef.

Posters and sketches

Historical posters included the promotional posters published by GBRMPA and by the QGTB. Those posters revealed little of the exact nature of particular reefs or reef species. However, they did represent cultural constructions of the Great Barrier Reef at particular periods in the history of the region: particularly during the period of rapid tourism expansion from around 1930. Therefore, reproductions of posters were collected in my research; they convey popular perceptions and attitudes towards the Great Barrier Reef. Some sketches were also collected as data. They included sketches found in the manuscript collections of the NLA and in the *Sailing Directions* of the lighthouse supply vessel, the *Cape Moreton*; the latter source revealed several physical changes in coral reefs, including the blasting of access tracks and channels. Sketches also included the illustrations produced by oral history informants, such as annotations made on printed maps, during qualitative interviews. One informant, for example, sketched the location and extent of the coral mining operation at Snapper Island reef. Some valuable sketch maps, identifying sections of reefs for which coral mining and coral collecting permits were issued, were found in the QSA.

Collection of oral data

My oral data collection took three forms: re-interrogation of existing oral histories, collection of original oral histories, and recording of oral fragments from other sources. The methods of data collection for each are discussed in turn below.

Pre-existing oral histories

Pre-existing oral histories about the Great Barrier Reef were gathered from three sources. First, the oral history recordings of the School of History, JCU, were obtained; they consisted of eighteen audio cassettes, produced in 1983, and a written summary of their main themes. That collection was originally created in order to investigate crown-of-thorns starfish (COTS) outbreaks in the Great Barrier Reef; however, those oral histories were re-interrogated for information about more general, historical changes in coral reefs, islands and marine wildlife species. Second, the extensive oral history collection of the NLA was searched for recordings relevant to the Great Barrier Reef; ten relevant recordings were interrogated for data about changes in the Great Barrier Reef and they revealed many perceptions of changes in the ecosystem, including many references to COTS infestations. Third, one private oral history recording was obtained, in which an individual had recorded an informal qualitative interview; it contained an account of the depletion of fish populations in parts of the Great Barrier Reef.

Original oral histories

Original oral histories formed a major part of my research. Initially, a list of key (expert) informants was created using purposive (convenience) sampling; those individuals were contacted to request an interview and access to historical photographs. Qualitative, semi-structured interviewing was used (with written consent), with audio recording and full transcription of the interviews following the conventions described by Gillham (2000) and the guidelines specified in the *Oral History Handbook* of the Oral History Association of Australia (South Australia Branch), compiled by Robertson (2000). Further informants were recruited using a snowballing technique; in total, 50 interviews were conducted and 47 of those were included in the final oral history collection, deposited at the GBRMPA Library in Townsville.

Anecdotes and comments

Other oral data included brief anecdotes and comments recorded throughout the course of the research. Those were collected in memo form and they served primarily to guide and focus the research process and to inform the qualitative interviews.

Data analysis

My data analysis involved the textual analysis of documentary materials and of oral history transcripts, using a system of coding, classification and grouping based on pre-defined categories (although other categories emerged during the process). I compiled a database of qualitative evidence for various coral reefs, islands and types of marine wildlife, although those primary categories were supplemented by others, including the various coastal locations, historical industries, environmental impacts, environmental management practices and perceptions of environmental change. Those analytical categories allowed the production of discrete historical vignettes that were built up into longer, more complex narratives about changes in coral reefs, islands and marine wildlife. Further analytical distinctions were made according to the type of material used, which also facilitated the cross-referencing of sources.

Visual representations, in most cases, were annotated and used to supplement documentary and oral data. In particular, the historical maps and sketches collected were interpreted in the light of other documentary records. For example, sketch maps of coral mining locations were matched with the archival records of the licences for those areas. Maps, sketches and photographs could also be dated if they illustrated activities for which surviving documentary records exist, or for which oral data is available. For example, a photograph depicting turtle hunting in the Fitzroy River was identified more precisely as a result of an oral history interview with one of the turtle fishers in the photograph.

Therefore, the analysis of documentary, oral and visual data together allowed the cross-referencing of sources and allowed a fuller historical narrative to be written; this was one advantage of the use of an array of qualitative methods in my research.

Summary

This chapter has outlined the methodology used in my research, including the sources of data available and the methods of data collection and analysis. My research involved the use of an array of diverse qualitative materials and a variety of interpretive techniques. However, textual analysis of archival documents and oral history transcripts was the main method used. Despite some limitations, as described above, those materials provide abundant original evidence of some environmental changes in the Great Barrier Reef for the period 1860–1960. My research included an evaluation of the use of qualitative sources in environmental history research for a coastal and marine environment, so the limitations of each type of material were made explicit. Those various qualitative materials have been used to construct the environmental history narrative that follows. However, before telling that story, some contextual material is presented first, in the next two chapters.

Chapter 3

The natural context of changes in the Great Barrier Reef

Introduction

Since European settlement, many historical changes have occurred in the environments and ecosystems of the Great Barrier Reef, particularly in the more accessible reefs and islands of the Cairns, Townsville and Capricorn-Bunker areas. Those changes occurred at many geographical scales, ranging from widespread effects (such as the impact of deteriorating water quality) to more localised impacts (such as the destruction caused by military target practice and various forms of mining). Furthermore, those changes occurred at various temporal scales, including long-term changes (such as the cumulative impacts of coral and shell collecting) and short-term changes (such as damage due to tropical cyclones). Indeed, classifying particular changes in the Great Barrier Reef is a difficult task because reefs are highly dynamic systems that are characterised by almost constant change at these various scales. Therefore, the narrative presented in this book documents changes in the Great Barrier Reef that have resulted from both natural and human causes – or from a combination of both – but where one impact ends and another begins is difficult to delineate precisely. Although, in this book, changes in the Great Barrier Reef are often categorised according to causal factor (and sometimes by time period), it is important to recognise that changes in the ecosystem have occurred as a result of multiple, combined impacts whose effects have varied geographically and temporally.

Given that complexity, it is important to understand the historical changes in the Great Barrier Reef against a background of ongoing, natural processes in the ecosystem. Some of those natural processes are powerful forces that may cause substantial degradation of particular coral reefs – or may even lead to widespread deterioration of parts of the ecosystem over time. At the same time, however, some parts of the ecosystem may display a remarkable capacity for regeneration and recovery from natural disturbances, with the result that the condition of the Great Barrier Reef at any given time represents a complex mosaic of coral reefs – and their associated habitats – in varying stages of degradation, decline and recovery. So it is important to be cautious in interpreting the effects of human activities on the Great Barrier Reef: those activities have occurred in a highly patchy and dynamic ecosystem in which specific impacts may be very difficult to discern and attribute with confidence.

To aid the understanding of historical changes in the Great Barrier Reef, therefore, this chapter begins with an outline of the large-scale, geomorphological changes that have occurred in the region as a result of the Holocene evolution of the continental shelf, which has involved profound changes in sea level and in sedimentation patterns. This outline represents a morphogenic approach to the evolution of the Great Barrier Reef and provides a context for the narratives of anthropogenic, historical changes that follow. Such an approach indicates that anthropogenic changes are influenced by geomorphological factors that have made some reefs highly vulnerable to degradation and decline. Another natural source of change in the Great Barrier Reef – the tropical cyclones that occur frequently in the region – is considered briefly next. Within that natural context, the accounts of human activities (such as the early European fisheries, coral mining, coral collecting and shell collecting) that follow suggest that some already-vulnerable reefs were intensively exploited, over a long period of time. In addition to those sustained activities, other, more sporadic human activities (such as military target practice and channel blasting) have also been superimposed, contributing to the creation of complex patterns of vulnerability, degradation and decline of some parts of the Great Barrier Reef. Consequently, some parts of the ecosystem were probably far from pristine at the time of the formation of the GBRMP, in 1975.

The Holocene evolution of the continental shelf

The Great Barrier Reef is a dynamic ecosystem characterised by almost continuous change at a wide range of geographical and temporal scales; those changes are largely controlled by geomorphological and climatic factors (Hopley, 1982, 1994; Hopley et al., 2007). Coral growth in the region has varied in rate and extent during the Holocene epoch, accelerating as hydro-isostatic processes adjusted sea level and as sea surface temperature (SST) increased, but also inhibited by mechanical erosion and ecological processes, such as bioerosion, which denude coral. Therefore, coral reefs represent the outcome of a balance between constructive and destructive processes, and they are typically patchwork assemblages of living, dying and dead corals (and coralline algae). However, such spatial and temporal patchiness does not necessarily indicate the overall decline of coral reefs. In addition to the natural variability that occurs on coral reefs due to the balance between coral (and algal) growth and bioerosion, coral reefs are also subjected to periodic changes as a result of variations in geomorphological, climatic, meteorological and biological factors. One of the most significant of those factors is the Holocene evolution of the continental shelf in the region, including variations in sea level and sedimentation patterns, which has profound implications for the vulnerability of the Great Barrier Reef to other impacts and which represents an important part of the large-scale context in which other, historical changes in the ecosystem have occurred.

The modern coral reefs of the Great Barrier Reef evolved during the Holocene: an interglacial epoch of dramatic changes in sea level and in the position of the eastern Australian coastline. The north-eastern Australian continental shelf experienced a rise in sea level of over 100 metres to around its present level, as a result of glacial ice sheet melting and the isostatic response of the continental shelf to water load, although the precise details of sea level history in the region are complex and still debated (Hopley et al., 2007). As sea level rose during the early Holocene, the north-eastern Australian coastline migrated laterally to its present position as the continental shelf was inundated. The modern Great Barrier Reef evolved on the newly-formed continental shelf, but in varying geomorphological conditions that controlled the rate and location of coral reef development. In particular, variations in the dominant geomorphological controls on the growth of coral reefs – sea level and sedimentation – have created, in different places and at different times, both favourable and hostile conditions for reef development. As a result of these variations, over around 6,000 years, some reefs have been brought close to thresholds of decline and may have experienced deterioration for geomorphological reasons. Conversely, other reefs – especially some offshore reefs – have flourished throughout the Holocene since they lie outside the region of particular vulnerability to geomorphologically-controlled decline.

Hopley (1994) has argued that, as sea level rose along the eastern Australian coast during the early Holocene, the corals of the Great Barrier Reef were able to grow upwards at a similar rate, forming a barrier within which further reef development was possible. However, during the early Holocene period of steady sea level rise, the position of the shoreline migrated westwards, resulting in the displacement of the Holocene sediment deposition zone across the shelf, with the result that sediment did not accumulate to any great depth in any particular area. However, once modern sea level was reached, around 6,000 years ago, the pattern of sediment discharge from the mainland to the Great Barrier Reef lagoon – which was dynamic during the period when the coastline was migrating – subsequently stabilised. The result was that the zone of terrestrial influence became static in its present position and sedimentation became concentrated in the newly-formed nearshore zone. This led to the formation of an inshore mud-silt wedge in the Great Barrier Reef lagoon, composed of sediments up to 15 metres in depth, as demonstrated by seismic surveys (Hopley, 1994, pp318–19; Hopley et al., 2007).

The stability of sea level and sedimentation patterns after around 6,000 years ago initiated considerable changes in the reefs of the Great Barrier Reef. Geomorphological descriptions of the Great Barrier Reef are based on a classification of different reef types according to their morphology and stage of development, and various classification schemes exist (Hopley et al., 1989). Although some reef types, such as ribbon reefs and fringing reefs, do not fit strictly within a developmental framework, in geomorphological terms most of the reefs in the Great Barrier Reef can be classified as juvenile, mature or senile. Hopley (1994, pp325–6) has argued that the progression of coral reefs from the juvenile stage to the mature and senile stages occurs under conditions of eustatic and isostatic stability: periods when tectonic movements of the continental crust are limited and sea level is relatively constant. During those periods of stability, coral reefs grow upwards to reach sea level and subsequently develop horizontally. During the mature phase of reef development, lagoons are formed; in the senile stage, the growth of live coral is restricted to the edge of the reef, while sediment infill occurs on the reef flat. Therefore, this model suggests that, in periods of tectonic and sea level stability, the deterioration of coral reefs occurs naturally as reefs progress through the juvenile and mature stages to reach a condition of senility.

Hopley (1994, p319) acknowledged that the Great Barrier Reef has experienced tectonic and eustatic stability since around 6,000 years ago, when many of its coral reefs commenced the transition to maturity and senility. Therefore, he suggested, many reefs of the Great Barrier Reef have declined from a juvenile state, in which rapid vertical coral growth took place, and instead have become characterised by sediment-covered reef flats, extensive patches of dead coral and comparatively small margins of live coral growth. Using this model, the coral reefs of the Great Barrier Reef can be divided into four main groups:

(a) fringing and nearshore reefs that have been severely impacted by sedimentation, displaying high mortality and limited recovery;
(b) fringing and nearshore reefs that have been significantly impacted by sedimentation, but that display ecological change, spatial patchiness in mortality, and some capacity to recover;
(c) mid-shelf reefs that have experienced terrestrial impacts, resulting in increased vulnerability to bioerosion and displaying increased rates of coral rubble formation; and
(d) offshore reefs which may have been affected by terrestrial influences, but for which degradation is only detectable using geochemical analysis techniques.

This framework is illustrated in Table 3.1, together with some examples of coral reefs that display those characteristics.

The significance of this morphogenic approach lies in the possibility that many reefs – particularly some fringing and nearshore reefs – may now be characterised by extreme vulnerability to other impacts; they may exist naturally close to ecological thresholds beyond which recovery from further degradation is very difficult, if not impossible. Therefore, for geomorphological reasons, along the Queensland coast – and particularly in the nearshore zone – the thresholds that determine whether or not coral reefs can recover from stresses are now likely to be relatively easy to cross. As a consequence, historical human impacts could easily have exceeded critical ecological thresholds, especially on some vulnerable, nearshore reefs. For those reefs, the effects of even comparatively slight historical human activities may have caused ecological phase shifts. Furthermore, once coral growth has been inhibited, for whatever reason, it may

Table 3.1 Main types of coral reef classified by geomorphological characteristics

Type of reef	Characteristics of reef	Examples
Severely impacted reefs	Reefs influenced by a nearshore mud-silt wedge Conditions of continuous turbidity Mortality of corals with little or no recovery Complete collapse of reef ecosystems	Stone Island reef Goold Island reef Alexandra Reef
Significantly impacted reefs	Patchy reefs High variability in coral mortality and recovery Ecological change in coral reef ecosystems Selection of sediment-resistant species Survival of impact-resistant forms of coral Increased prevalence of soft corals	Palm Island reefs Halifax Bay reefs Middle Island reefs Cape Tribulation reefs
Moderately impacted reefs	Mid-shelf reefs with limited terrestrial influence High cover of living corals Recovery of corals from severe impacts Increased coral skeleton porosity Increased bioerosion of corals	Holbourne Island reef
Slightly impacted reefs	Offshore reefs with slight terrestrial influence Healthy, resilient corals Terrestrial influence not visible Geochemical analysis required to reveal impacts	Wallaby Reef Kangaroo Reef

Source: Based on information provided by D. Hopley, personal communication, 20 October 2003

be much harder for the recovery of reefs to take place; in other words, it may now be much easier to 'turn-off' than to 'turn-on' coral growth in many parts of the Great Barrier Reef. Moreover, this approach suggests that, once a coral reef is established, it may withstand fairly poor environmental conditions; but, once a critical ecological threshold is crossed and a phase shift occurs, then the quality of those conditions may need to be improved significantly for recovery of that reef to occur. The degradation of parts of the Great Barrier Reef – for both natural and anthropogenic reasons – therefore increases the vulnerability of the ecosystem to other disturbances, even the effects of single events such as tropical cyclones or floods.

The deterioration of the coral reefs within 20 or 25 kilometres of the Queensland coast, therefore, has probably occurred as the impacts of sediment and nutrient run-off from the mainland have accelerated the natural tendency of those reefs to reach a stage of geomorphological senility (Hopley, 1994). The naturally-occurring mud-silt wedge, which may extend for several kilometres offshore, encroaches on some nearshore reefs (as in Halifax Bay) and also on the fringing reefs of some nearshore continental islands. In those locations, the impacts of terrigenous sediment deposition on coral reefs have been exacerbated by high water velocities: this is because high rates of sediment accumulation have coincided roughly with the effective wave zone, which has caused re-suspension of the sediments, representing a natural process of deteriorating water quality over the last 6,000 years (Hopley, 1994). Again, a geomorphological perspective suggests that caution is required in interpreting the deterioration of nearshore coral reefs in the Great Barrier Reef: it is not possible to attribute that deterioration solely to anthropogenic influences, as there are also geomorphological reasons for the decline in water quality in the region. Indeed, from a geomorphological perspective, the deterioration of the nearshore coral reefs of the Great Barrier Reef is almost inevitable, since the control exerted by sea level no longer allows juvenile reefs to form.

This morphogenic perspective provides a natural, large-scale context for the accounts of historical changes in the Great Barrier Reef that follow. Those recent changes should be interpreted against a background of the high vulnerability of fringing and nearshore regions reefs as a result of their Holocene evolution. The anthropogenic activities that are described in subsequent chapters – including coral mining and coral collecting – have affected some coral reefs that already had limited capacity to recover from environmental stresses. In some cases, the impacts of those activities probably caused the complete mortality of parts of some reefs; the degradation of the reefs at Goold Island, Kings Reef and Alexandra Reef may have occurred in this way. Yet some anomalies exist in this framework: Middle Reef, near Townsville, for instance, appears to display an unusual degree of resistance to mortality, despite experiencing highly turbid water conditions; on that reef, one oral history informant stated, an 'absolutely amazing amount of coral' was found.[1] The reefs of Halifax Bay, similarly, contain some apparently resilient reefs, possibly as a result of their more stable foundations on Pleistocene gravels. Nevertheless, this model of geomorphological controls on the Holocene evolution of the Great Barrier Reef provides a valuable means of interpreting the impacts of human activities and changes in vulnerable coral reefs.

Tropical cyclone damage to coral reefs

During the Holocene evolution of the Great Barrier Reef, coral reefs have also experienced geomorphological changes due to tropical cyclone-related wave action, including abrasion as coral fragments and other debris are thrown against coral colonies. The patterns of occurrence and severity of tropical cyclones in the Great Barrier Reef have been reconstructed by several authors. Puotinen et al. (1997) documented the historical frequency and paths of some cyclones in the Great Barrier Reef, demonstrating the recurrent nature of those storms in the region. Nott and Hayne (2001) and Nott (2003) have reconstructed the severity of tropical cyclones in the region, showing that some of those storms could be exceptionally destructive. Tropical cyclones, therefore, represent another environmental factor that has caused changes in coral reefs; in particular, affecting fragile corals such as *Acropora* species, although other species are also susceptible to storm damage. Documentary sources refer to noteworthy tropical cyclones at Green Island (1858), Cooktown (27 January 1899 and 4–5 March 1899) and Low Isles (March 1911) (Almond, 1899, p1025; Mackay, 1911, p1187).[2] While travelling in the Great Barrier Reef, Agassiz (1898, p115) referred to Saville-Kent's observation of 'the wreckage of the fringing reef by a hurricane at Saddleback Island'. For some locations, a time series of tropical cyclones has been reconstructed; for example, Loch (1991, p5) showed that severe tropical cyclones affected Michaelmas Cay in March 1878, January 1906, March 1911, February 1920, February 1927, March 1934 and 1948, in addition to many smaller storms. Another record of this type of naturally-occurring damage stated that the jetty at Green Island was destroyed by a tropical cyclone in 1946 and was reconstructed by the Cairns Harbour Board.[3]

The GBRC expedition to the Great Barrier Reef in 1936 described other instances of damage to corals that were attributed to tropical cyclones. For example, observers on that expedition stated that Mackay Cay was 'severely damaged' by a tropical cyclone in 1934 and, at the reef between Ingram and Beanley Islands, the same report indicated that 'the sandy surface of this reef was caused by the destruction, through a cyclone, of a former cay' (Steers, 1938, pp70, 84). The observers found that Night Island had been devastated by a tropical cyclone within the preceding 20 years. Since that event, Steers (1938, pp94–5) stated:

> To the north-west of the reef the mangrove mud has spread, and seems to have killed much of the reef. Incidentally, much of the whole area covered by the mangroves was sandy; there was also abundant *Thalassia*. [...] Whilst the upper surface of the reef seems to be largely moribund, the general appearance of much of the cay and mangrove area is one of recovery and rejuvenation after a severe blow.

Moreover, the impacts of tropical cyclone damage were not restricted to the northern Great Barrier Reef. For instance, Steers (1938, p54) described the major transformation of Lady Elliot Island reef that he attributed to the tropical cyclone of March 1936, which 'appears to have been responsible for building the outer ridge' of the reef.

Oral history sources provide additional details of the impacts of tropical cyclones on coral reefs. One informant, a geomorphologist, recalled the visible effects of the storm that struck the Bowen area in 1918, affecting the coral reefs at Stone and Holbourne Islands. At Stone Island, he reported that almost no coral had survived, even where coral communities would now be expected to be found.[4] Considerable damage also occurred at Holbourne Island when the reef moat was breached by a storm, leading to a dramatic change in water level.[5] The informant stated:

> A cyclone hit [Holbourne Island] in 1918; the island prior to this – or the fringing reef – had a shingle ridge around the outer edge of the reef, which moated the water at low tide. Within this moat, there was quite good [...] living coral. What happened during the cyclone was that the shingle rampart was breached; water levels became much lower on the reef flat and a lot of the living corals just died off. They are still there; they are high micro-atolls and you can see – about 30 or 35 centimetres below that level – where coral has grown since.[6]

The recovery of Holbourne Island reef – in contrast to the reef at Stone Island – was attributed, by that informant, to reduced sedimentation at Holbourne Island, which is further offshore and more distant from terrestrial impacts than is Stone Island.

Oral history evidence indicates that tropical cyclone damage to corals has been witnessed by many observers, including catastrophic reductions in coral cover. One informant, a coral reef scientist, reported seeing changes in coral reefs: 'from incredibly rich coral communities with 50 to 75 per cent coral cover down to less than five per cent coral cover'. However, the same informant described the rapid recovery of offshore coral reefs from tropical cyclones, stating:

> You just get huge recruitment and rapid growth of *Acroporas*. Going back in five years' time after total devastation will show you what is apparently quite a healthy reef, although, if you look closely, you'll see most of the corals are less than half a metre in diameter. So you can get rapid recovery in exposed, high-energy situations.[7]

In contrast to offshore reefs, fringing reefs are particularly vulnerable to the effects of tropical cyclones. The informant stated that 'a good cyclone reduces them to rubble with virtually no coral cover'.[8] As a result, historical changes in fringing reefs can be over-written by the influence of successive storms, resulting in changes to the structure of those reefs as coral rubble and larger coral pieces are transported by wave action.

Many informants recalled the effects of particular tropical cyclones on specific reefs. One informant, a sugar cane cutter and recreational fisher, referred to the storm that struck Port Douglas in 1911; he also described the tropical cyclone that occurred at Cape Tribulation on 12 March 1934.[9] Another informant, a shell collector, described the extent of the damage at Orpheus Island reef, stating that:

> I was in my early teens when we visited Orpheus Island, in the Palm [Island] group, and saw first-hand what destruction the power of a tropical cyclone can create: huge banks of broken coral metres deep cast high into the vegetation in drifts. By sifting through this coral, we found lots of spectacular shells we had only seen illustrated in Joyce Allen's *Australian Shells*.[10]

Another informant, a coral reef scientist, witnessed tropical cyclone damage at Heron Island, when the disturbance came from an unusual direction and affected corals that had not adapted to cyclone conditions (see also Bennett, 1971, p25).[11] Many other oral history accounts describe the impacts of tropical cyclones, although much of that evidence refers to the period after 1970, which is relatively well-documented by other sources.[12] Nonetheless, there is abundant evidence that many coral reefs have been damaged periodically by tropical cyclones, although offshore reefs have generally recovered more rapidly from storm damage; in contrast, fringing and nearshore reefs have experienced slower recovery rates or – as at Stone Island – no recovery has apparently occurred.

Summary

This chapter has presented a brief overview of some of the major natural factors that have influenced the condition of the Great Barrier Reef: the geomorphological evolution of the continental shelf during the Holocene epoch, and the effect of the tropical cyclones that are a recurrent feature of the region. Those natural processes and events have been described here in order to provide a context for the accounts of historical human impacts on the ecosystem that follow. Due to the Holocene evolution of the continental shelf, changing patterns of sea level and sedimentation mean that some of the coral reefs of the Great Barrier Reef – especially those in the nearshore zone – are now found in a geomorphologically senile condition and are therefore highly vulnerable to other impacts. Superimposed upon that natural context of deterioration are the impacts of tropical cyclones on individual coral reefs, which may be severe, although (in the absence of other significant pressures) many reefs nevertheless display a remarkable capacity to recover from the effects of those storms. This natural context – one of variable and patchy deterioration and recovery – means that the accounts of human activities that follow must be interpreted with care. Yet it also suggests that some already-vulnerable reefs – ones that naturally exist close to critical ecological thresholds – may have been the same ones that have been intensively exploited by humans, sometimes over long periods of time. Those superimposed patterns of both natural and human impacts have led to the creation of complex patterns of vulnerability, degradation and decline of some parts of the Great Barrier Reef, and they indicate that some parts of the ecosystem were probably far from pristine at the time of the formation of the GBRMP in 1975.

Notes

1. Oral History Cassette (OHC) 35, 20 October 2003, *Changes in the Great Barrier Reef since European Settlement*, Oral History Collection, School of Tropical Environment Studies and Geography, JCU, October 2002–December 2003, pp10–11.
2. GBRMPA, 'Green Island economic study: summary report, October 1979', Economic Associates Australia, Economic and Management Consultants, 1979, Appendix A: history of Green Island and its reef, SRS5416/1 Item 434, QSA.

3 GBRMPA, 'Green Island economic study: summary report, October 1979', Economic Associates Australia, Economic and Management Consultants, 1979, Appendix A: history of Green Island and its reef, SRS5416/1 Item 434, QSA.
4 OHC 35, 20 October 2003.
5 OHC 35, 20 October 2003.
6 OHC 35, 20 October 2003.
7 OHC 20, 9 September 2003.
8 OHC 20, 9 September 2003.
9 OHC 17, 2 September 2003.
10 Anonymous, 'Recollections of the reef', Changes in the Great Barrier Reef since European Settlement, Oral History Collection, School of TESAG, JCU, September 2003, pp1–2.
11 OHC 4, 14 January 2003.
12 Additional details are found in OHC 1, 30 October 2002; OHC 5, 11 February 2003; OHC 6, 17 February 2003; OHC 16, 2 September 2003; OHC 18, 5 September 2003; OHC19, 9 September 2003; OHC 20, 9 September 2003; OHC 26, 17 September 2003.

Chapter 4

The spread of European settlement in coastal Queensland

Introduction

This chapter provides a brief overview of the history of European settlement on the coast of Queensland, based on secondary sources. In particular, I have used the histories written by Bolton (1981), Fitzgerald (1982, 1984), Reynolds (2003) and Evans (2007), and the Australian historical geography by Powell (1988), all of which describe European settlement in Queensland. This chapter also draws heavily on the works of historical geography by Griggs (1997, 1999a, 1999b, 2000, 2003, 2004, 2011). It situates the subsequent narrative of changes in the coral reefs, islands and marine wildlife of the Great Barrier Reef in the context of the northward expansion of European settlement on the Queensland coast and the establishment of a new pattern of land use: one based on the dominant activities of pastoralism, sugar cane farming, mining and tourism. Such a historical context is required because the decline of the Great Barrier Reef has been partly attributed to those terrestrial activities (see Chapter 1), and it is now widely acknowledged that some parts of the Great Barrier Reef have been significantly affected by terrestrial activities in the adjacent catchments (Williams, 2001; Williams et al., 2002; Furnas, 2003).

First, this chapter briefly outlines the context of the history of European settlement in coastal Queensland: the evolution of the Australasian continent, the formation of the Great Barrier Reef, Aboriginal occupation, pre-European contacts and the earliest European settlement in Australia. Next, I describe the European settlement of Queensland since the formation of its first colony, at Moreton Bay, with an emphasis on the introduction and expansion of sugar cane farming. However, other European activities – pastoralism, mining, timber-getting, *bêche-de-mer* harvesting and pearl-shell collecting – were also important during that period and are also mentioned. In addition to those activities, recent coastal development and the emergence of conservation concerns in Queensland are considered. This brief review suggests that significant impacts on some parts of the Great Barrier Reef occurred during the period of European settlement, although those impacts varied geographically and at different times.

The context of European settlement in coastal Queensland

European settlement in Queensland followed a long and complex history of changes in the natural and cultural environment. Archer et al. (1998) described many such changes in their account of Australian environmental history during the last 100 million years. Those changes include the evolution of the Australasian continent since the disintegration of the Gondwanaland supercontinent, the northward drift of the Australian tectonic plate, fluctuations in sea level, the formation of the Great Barrier Reef, the arrival of humans in Australia, the development of trading relationships between Indigenous Australians and neighbouring societies, and the earliest European settlement in Australia. Below, I discuss these changes briefly in order to provide a context for the history of European settlement in Queensland that follows. The account presented in this section also indicates the significance of the Great Barrier Reef in shaping the course of the European settlement of Queensland. Bolton (1981, p1) began his account of the history of north Queensland by referring to the Great Barrier Reef – acknowledging its danger to navigation – and Bowen and Bowen (2002) demonstrated that the Great Barrier Reef has been of critical importance in influencing the development of the colony of Queensland, both as a hazard and as a resource.

The history of the north-eastern coast of Australia commenced with the formation of the Australasian continent after the fragmentation of the Gondwanaland supercontinent, around 200 million years ago (Lunine, 1999, p95). The Australian tectonic plate moved northwards by the process of continental drift to its present position, in which a large part of the Queensland coast lies within the tropical zone with conditions suitable for the growth of coral reefs. During the continental drift of the Australian tectonic plate, terrestrial aridity increased and considerable changes occurred in the vegetation and biota of the continent. In addition, sea level fluctuated in response to alternating glacial and interglacial climatic regimes and, during the most recent 10,000 years of the Holocene epoch, ice sheet melting and isostatic adjustments of the Australian plate caused sea level to vary in north-eastern Australia until present sea level was reached, around 6,000 years ago, although the details of sea level history in the region are complex (Hopley, 1997; Veron, 2009). After sea level stabilised, the modern Great Barrier Reef formed on limestone foundations composed of the remains of Pleistocene reefs; hence, the modern Great Barrier Reef is a comparatively recent structure in geological terms (Bird, 1971; Hopley, 1982; Hopley et al., 2007) (see Chapter 3).

At several times during the evolution of the continent, lower sea levels existed between northern Australia and south-eastern Asia than occur at present, which allowed successive waves of human migration into Australia. Australia was populated by human societies that may have occupied the continent for more than 50,000 years; a complex pattern of Indigenous countries was created during that period (Horton, 1994; Veron, 2009). By the time of the earliest European settlement in Australia, the terrestrial environment had been modified by Aboriginal land management practices, including the manipulation of soils with yam sticks, the transformation of vegetation using fire and the hunting of fauna. Those impacts were considerable, although their nature and extent has been contested (Benson and Redpath, 1997; Choquenot and Bowman, 1998). Trade between Aboriginal clans was widespread and was organised along major river routes; contact between Indigenous Australian societies and other peoples, including Papuan, Cantonese and Macassan traders, was also extensive. As a result, the transformation of the Australian terrestrial environment by Indigenous societies was substantial in both its geographical extent and its duration (Fitzgerald, 1984; Johnston, 1988; Hill et al., 1999, 2000; Crowley and Garnett, 2000). The Great Barrier Reef was also used extensively by Indigenous Australians as a source of food, tools, ornaments and trading commodities; hunting of dugongs, for example, represented an important part of Indigenous Australian social and cultural life, as Marsh and Corkeron (1997) acknowledged, and the hunting of marine turtles was also culturally significant for coastal Indigenous Australians (McCarthy, 1955; James, 1962).

European contact with the Australian environment began with the early exploratory voyages made by Dutch, English, Spanish, Portuguese and French mariners. Bowen and Bowen (2002, pp14–15) discussed the evidence that Portuguese and Spanish sailors charted parts of the north-eastern coast of Australia in the years after the Portuguese settlement of Timor, in 1516, although that evidence is inconclusive because many Portuguese maps were lost in the Lisbon earthquake of 1755. In 1606, part of the eastern coast of the Gulf of Carpentaria was charted by the Dutch crew of the *Duyfken*; the Spanish navigator, Luis Vaez de Torres, sailed through Torres Strait in the same year. In 1616, a Dutch vessel, the *Eendracht*, reached the northern and western coasts of Australia during voyages from Europe to the East Indies, and Dutch ships later sailed to Batavia via northern Australia. Parts of the southern coast of Australia were also charted by Dutch mariners: in 1642, Abel Tasman reached southern Tasmania. In 1622, an English ship, the *Trial*, following the same route as Tasman, sighted Australia; later, in 1688, William Dampier reached north-western Australia. The French navigator, Louis-Antoine de Bougainville, sailed through the Coral Sea in 1768 and came within sight of the Great Barrier Reef.

In 1770, the British navigator, James Cook, charted the eastern coast of Australia in *HMS Endeavour* and claimed possession of that land for the British Crown. After the declaration of independence by the North American colonies, the British Empire faced a penal crisis that was resolved by sending convicts to Australia. The first British settlers reached Australia in 1788, when the First Fleet, commanded by Arthur Phillip, arrived at Botany Bay with a population of around 1,400 people, consisting mostly of convicts, sailors and marines. The first European settlements were established at Sydney Cove, Parramatta and Norfolk Island. Small farming was established on plots of land occupied by emancipated convicts; subsequent settlement occurred at Hobart, in Van Diemen's Land (now Tasmania) in 1804, and also at Port Arthur. The separation of Van Diemen's Land from the colony of New South Wales took place in 1825 and, from 1824–1836, four other settlements were created: Moreton Bay (now Brisbane) in 1824, Swann River (now Perth) in 1829, Port Phillip (now Melbourne) in 1835 and Adelaide in 1836. Farming, grazing and gold mining took place in the hinterlands of those settlements, stimulating their economic growth and attracting new migrants.

While European settlement spread along the Australian coast, inland exploration also took place. Major expeditions included the journeys made by Oxley (1817), Sturt (1828), Mitchell (1835–1836 and 1844–1845), Eyre (1840–1841), Warburton (1872–1873), Leichhardt (1844–1845), Kennedy (1848), Burke and Wills (1860–1861), Stuart (1861–1862) and Giles (1876). Those expeditions facilitated the movement of European pastoralists and squatters inland, although European settlers encountered various forms of Aboriginal resistance, as several authors have narrated (Bolton, 1963; Loos, 1982; Reynolds, 1982, 1987, 2003, pp vi, 11–12; Birtles, 1997, p394). Nevertheless, the period from 1850–1889 was characterised by rapid economic development in the Australian colonies, stimulated by exports of wool and discoveries of gold. Boom towns, such as Ballarat and Bendigo, prospered as gold fields attracted new European migrants. Further immigration also encouraged the growth of the major cities, especially Sydney and Melbourne, and the establishment of new ports such as those at Rockhampton and Townsville. Consequently, by the last decade of the nineteenth century, large areas of Australia had been settled by Europeans; pastoralism, agriculture and mining were expanding, and the population was approaching four million people. A period of economic depression and drought from 1890–1906 marked the end of that period of rapid European settlement in Australia.

European settlement in coastal Queensland

European settlement in the area that would later become Queensland began in the south-east and spread rapidly northwards and inland. The first settlement, known as the Moreton Bay colony, was established at Redcliffe in 1824; it was initially a convict settlement but, by 1840, free settlement had also begun in the colony. The site chosen at Redcliffe was advantageous because of the availability of safe anchorage and pastoral opportunities in its hinterland, although it lacked adequate fresh water and, in the following year, the settlement was transferred to a more favourable site where modern Brisbane stands (Fitzgerald, 1982, p74). As occurred elsewhere in Australia, pastoral occupation took place in the region surrounding the settlement; from 1840, the migration of pastoralists northwards in search of new grazing land was extremely rapid. As early as 1842, most of the Darling Downs had been claimed by pastoralists; soon afterwards, in 1847, a town settlement became necessary at Port Curtis (now Gladstone) (Fitzgerald,

1982, p95). Pastoral expansion continued and, in 1858, a third port was founded at Rockhampton. Shortly after the separation of Queensland from New South Wales, in 1859, a further coastal settlement was created at Port Denison (now Bowen).

Bolton (1981, p10) has argued that pastoral expansion was strongly related to exploration during that period. Exploration, including the expeditions by Kennedy (in 1848) and by Dalrymple (in 1859), provided an indication of available resources – in particular, identifying good pastoral areas – and squatters occupied land soon after its earliest exploration by Europeans. Frontier areas were also settled by new immigrants from Europe, with the encouragement of the New South Wales Government. Indeed, the movement of the pastoral frontier seemed so relentless that, in 1860, Sir George Ferguson Bowen (1889, p193), the inaugural Governor of Queensland, wrote:

> There is something almost sublime in the steady, silent flow of pastoral occupation over north-eastern Australia. It resembles the rise of the tide, or some other operation of nature, rather than the work of man [sic].

As a result of the rapid occupation of land, Bowen (1889, p193) stated that 'at the close of every year, we find that the margin of Christianity and civilisation has been pushed forward by some two hundred miles'.

The pastoral areas opened by European exploration required supplies of water, especially from the Burdekin, Fitzroy and Herbert Rivers; the location of rivers also determined the availability of fresh water for new settlements. During the European settlement of Queensland, therefore, many coastal ports were established adjacent to major rivers and were used by the Queensland Royal Mail Line steamers, which sailed monthly between Brisbane and London using the Inner Passage through the Great Barrier Reef, and whose operations were subsidised by the Queensland Government. On their return journeys, those ships brought new migrants to Queensland. In contrast, the development of terrestrial means of transport was slow, being hindered by the Great Dividing Range, and initially few roads and railways were constructed. Therefore, the charting of safe passages through the Great Barrier Reef was essential for the early development of Queensland. Extensive hydrographic surveying occurred during the voyages of Flinders (1802), Bunker (1803), Jeffreys (1815), King (1819), Oxley (1823), Wickham (1839 and 1845–1846), Stokes (1841), Blackwood (1843 and 1844–1845), Yule (1844–1845), Stanley (1848), Holthouse (1976) and Gill (1988). Those surveys informed publications for mariners, such as *The Australia Directory*, which in turn enabled more reliable shipping in the Great Barrier Reef. Surveying vessels also carried naturalists aboard, including Jukes and MacGillivray aboard the *Fly* and Huxley and MacGillivray aboard the *Rattlesnake*, who documented the voyages in their journals and collected scientific data.

The development of ports encouraged the expansion of many industries in coastal Queensland. In the northern part of the colony, coastal areas were pioneered by *bêche-de-mer* fishers and cedar-cutters. During the second half of the nineteenth century, *bêche-de-mer* was harvested from coral reefs in the Great Barrier Reef, processed at small curing stations and exported to south-eastern Asia. From 1874, timber-getters cut the forests of red cedar that were found on the Queensland coast between Cardwell and Cooktown. By 1880, merchants such as Burns, Philp and Company were trading in several tropical products, including *bêche-de-mer*, timber and copra (Bolton, 1981). In addition to those industries, gold mining took place at many locations, including the Palmer, Hodgkinson, Charters Towers and Ravenswood goldfields; the discovery of gold was responsible for the rapid growth of European settlements such as Cooktown and Charters Towers. However, pastoralism remained crucial to the Queensland economy throughout that period. From 1860–1900, the pastoral industry prospered as virtually all land suitable for grazing was taken up in leases; growth was also stimulated when frozen meat began to be exported to Britain during the 1870s (Bolton, 1981).

In addition to pastoralism, from the mid-1860s, agriculture became significant for the economic development of Queensland. Early attempts at cotton cultivation were short-lived; government subsidies for cotton farmers in the 1860s stimulated agricultural expansion, and cotton exports were successful during the period of the American Civil War, but the industry declined shortly afterwards. In contrast, sugar cane cultivation was more successful: the initial expansion of sugar cane cultivation, from 1864–1884, is shown in Figure 4.1(a). Sugar cane was first grown in Queensland, in the mid-1860s, in the Maryborough, Brisbane and Beenleigh districts, and rapid expansion took place in the sugar industry between the late 1860s and 1884 (Griggs, 1999b, 2011). By the 1880s, sugar cane farming had become prominent in the Queensland economy and contributed to the growth of the settlements at Mackay, Bundaberg, Maryborough, Geraldton (now Innisfail) and Cairns. The production of sugar cane was made more economic by the use of indentured Melanesian labourers; but, after 1884, a surplus of sugar derived from European sugar beet on the world market and the opposition, by the Queensland Government, to the recruitment and employment of Melanesian workers led to a contraction in the Queensland sugar industry in the late 1880s. After 1892, when the decision to restrict the use of indentured labourers had been reversed and the world sugar price had increased, confidence was restored in the sugar industry.

By 1900, sugar production had exceeded the demands of the colony, although an overall rapid expansion in the cultivated area of sugar cane continued to take place, as Figure 4.1(a) shows, and exports of sugar commenced. Those increases in sugar cane acreage and sugar yields were obtained as a result of the creation of new sugar cane fields, the expansion of production on existing sugar cane land and the increasing adoption of scientific methods in sugar cane farming. Griggs (2003, 2004, 2007) has shown that the expansion of sugar cane land between 1865 and 1900 resulted in some severe environmental impacts, including the complete deforestation of land to create farmland and the cutting of timber to provide sugar mills with a source of fuel. Alongside those environmental changes,

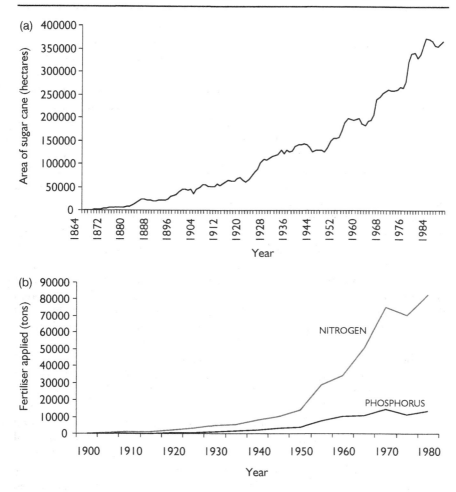

Figure 4.1 (a) Area of sugar cane grown in Queensland, 1864–1988; (b) Fertiliser application in the Great Barrier Reef Catchment Area (GBRCA), 1900–1980. Source: Based on information provided in *SCQ*, 1864–1900, *SSQ*, 1901–1915, and *Australian Yearbook*, 1916–1988, compiled by P. Griggs; Pulsford (1996, pp16, 22); Furnas (2003, p159)

the sugar industry was restructured: plantation production was replaced by the small cane farming system. In 1901, the Commonwealth Government legislated to replace Melanesian workers after 1906 with white labourers only in the Queensland sugar industry (Griggs, 1997, 1999a, 1999b, 2000).

By 1910, as Figure 4.1(b) illustrates, the application of nitrogenous fertiliser by sugar cane farmers had commenced in the GBRCA, following the outcome of soil analyses that were conducted by the Queensland Bureau of Sugar Experimental Stations (QBSES). As a result of those experiments, the QBSES succeeded in calculating the correct fertiliser application rates for sugar cane. Keating et al.

(1997) demonstrated that application rates of nitrogenous fertiliser by Australian sugar cane farmers increased since around 1910, with a very substantial increase since 1950; that large increase in nitrogen application is shown in Figure 4.1(b), which was accompanied by an increase in sugar cane yields (see also Garside et al., 1997). However, the rates of nitrogen application increased more rapidly than sugar cane yields, with the over-application of nitrogen in some cane-producing districts, with the result that the levels of nutrient runoff to the Great Barrier Reef increased. Figure 4.1(b) also shows the concurrent, although smaller, increase in phosphorus application in Queensland since around 1920. Phosphorus is another essential nutrient for sugar cane growth; and, without the use of artificial fertilisers, soil phosphorus levels are generally low in uncultivated or new sugar cane areas (Wood et al., 1997).

The soil analyses carried out by the QBSES showed that, in addition to nitrogenous fertiliser, the acidic soils of the northern coastal districts – including the Mossman, Cairns, Innisfail and Tully districts – required agricultural lime. However, prior to 1945, the Mossman, Cairns, Innisfail, Ingham and Mackay districts had only limited terrestrial sources of lime besides the inland sources at Chillagoe and Ambrose, near Mount Larcom.[1] In 1915, the QBSES reported that efforts were being made to produce agricultural lime by pulverising coral from the Great Barrier Reef; agricultural lime manufactured from coral had the advantages of being comparatively cheap and chemically pure. The following year, the QBSES found that interest in coral sand and coral lime was high among sugar cane farmers, and that pulverising machines were already available on the market; hence the QBSES advised farmers to use coral lime, chemical fertilisers and green manures. By 1920, coral lime was being applied in the Mossman, Goondi, Mourilyan and South Johnstone areas at a cost of £3 per ton for coral sand and £4 per ton for burnt coral lime (Scriven, 1915, 1916, 1922). The removal of coral from the Great Barrier Reef to produce agricultural and industrial lime is one of the activities described in the narrative that follows (Chapter 11).

By 1914, the modern pattern of European occupation in coastal Queensland and the dominant industries of grazing, sugar cane farming and mining had been established. Closer settlement and further land clearance took place after the First World War as additional lands were allocated to returning soldiers; for example, lands at El Arish, near Tully, were occupied by discharged soldiers. Other newly-opened lands included the rainforests of the Atherton-Evelyn Tableland, which were cleared to open new grazing pastures, for maize cultivation and for dairying (Birtles, 1988). The continued expansion of sugar cane cultivation in the two subsequent decades, shown in Figure 4.1(a), was accompanied by increasing transformations of the Queensland environment: additional land clearance; increases in fertiliser, insecticide and pesticide application rates; and ecological changes. For example, in 1935, the cane toad (*Bufo marinus*) was introduced in Queensland at the recommendation of the QBSES in an effort to combat cane grub outbreaks, with severe ecological consequences (Griggs, 2005, 2007).

In general, European settlement in Queensland was associated with increased soil erosion as cycles of pastoral and agricultural expansion were followed by drought, leading to the degradation of large land areas (Powell, 1988). Land degradation was exacerbated by forest clearance and by the construction of large-scale irrigation schemes. It was also due to the fact that, once the most suitable sugar cane lands had been cultivated, farmers increasingly resorted to the use of sloping land. Overall, the years 1880–1935 were characterised by high rates of soil erosion in the GBRCA. By 1939, soil erosion was acknowledged to be severe in almost every cane-producing district of Queensland, not least because of the cultivation of sloping land, but soil erosion was also exacerbated by the destruction of riparian vegetation in sugar cane districts. The problem of soil erosion from sugar cane lands remained severe before 1945, prior to the introduction of conservation tillage and contour farming practices. Consequently, substantial terrestrial runoff of sediment and nutrients to the nearshore waters of the Great Barrier Reef occurred (Courtenay, 1978; Crossland et al., 1997; Griggs, 2005, 2006, 2007, 2011).

By 1939, other extensive alterations of the Queensland coastal environment had occurred: examples include the depletion resulting from the European marine fisheries of that period, including the commercial *bêche-de-mer*, pearl-shell, dugong and turtle fisheries; the transformation of some islands as a result of guano and rock phosphate mining; the establishment of coconut palm plantations; and the construction of tourist resorts. Some of those activities, such as the operation of the turtle soup factories at North West and Heron Islands during the 1930s, occurred intensively until localised depletion of the marine resources caused production to cease, as described in the narrative that follows (Chapter 6). On the adjacent coastal land, insecticides and pesticides (including DDT, atrazine and diuron) were increasingly used, which in turn degraded water quality in the Great Barrier Reef (Crossland et al., 1997; Keating et al., 1997). In addition, the introduction of irrigation in the Queensland sugar cane industry, combined with inadequate drainage of farmland, contributed to waterlogging, a decline in soil fertility and enhanced nutrient and sediment runoff (Meyer, 1997). Therefore, while the modern form of the Queensland economy was established during the period 1860–1900, the subsequent period until 1940 involved more intensive exploitation of both terrestrial and marine resources and, by that year, the impacts of terrestrial activities on the Great Barrier Reef had become substantial.

After 1945, the Queensland coastal environment continued to experience modifications as a result of many human activities. More extensive environmental degradation occurred in coastal Queensland for several reasons: extensive drainage of swamps to create land for sugar cane cultivation; the destruction of wetland habitats; a substantial increase in soil erosion; and the continued growth of coastal settlements (Griggs, 2007, 2011). The impacts of human activities on the Great Barrier Reef have also been described by Lawrence et al. (2002); those activities include the expansion of the tourist facilities on Green, Hayman and Heron Islands, mangrove clearance to create urban and industrial land in the

Gladstone area, and the reclamation of coastal land in the vicinity of Cairns. After 1945, expansions took place in the grazing, sugar cane, tropical fruit, tobacco and mining industries. Increases in sugar cultivation, in particular, occurred as wartime shortages of fertiliser and labour were overcome and as the introduction of bulk-loading facilities facilitated sugar exports. The considerable expansion of sugar cane acreage in Queensland between 1952 and 1976, illustrated in Figure 4.1(a), occurred in response to rising world sugar prices and growing demand from Asian customers, especially in Japan, South Korea and China; that expansion led to further soil erosion and further inputs of sediment and nutrients to the nearshore waters of the Great Barrier Reef (Griggs, 2007, 2011).

Particular environmental degradation in coastal Queensland occurred during the 1960s and 1970s, when large areas of freshwater swamp were drained for sugar cane cultivation and many stream catchments were cleared to provide additional agricultural land. Those changes generated further soil erosion, substantially increasing sediment and nutrient runoff from the GBRCA (Arthington et al., 1997; Meyer, 1997; Furnas, 2003; Griggs, 2007, 2011). During that period, other resource exploitation was proposed in the Great Barrier Reef itself. Oil exploration occurred during the 1960s and permits for oil drilling in the Great Barrier Reef were issued by the Queensland Government (Fitzgerald, 1984; Hopley, 1989). However, by the late 1960s, the perceived extent of environmental exploitation and degradation in Queensland was prompting conservation concerns among the public (Bowen, 1994; Bowen and Bowen, 2002). In particular, a proposal by the Cairns District Canegrowers Association to mine coral from Ellison Reef, near Innisfail, generated unprecedented levels of environmental protest, which, together with public fears about the consequences of oil pollution in the Great Barrier Reef, eventually led to the formation of the GBRMP in 1975.

Since 1981, conservation in coastal Queensland has been facilitated by the designation of the GBRWHA and, in northern Queensland, of the adjacent Wet Tropics World Heritage Area (WTWHA). As a result, scientific monitoring of human impacts on the outstanding natural phenomena of those environments commenced and greater attention was paid to their management. In particular, the multiple impacts of terrestrial runoff, commercial and recreational fisheries, shipping and tourism (amongst other activities) are now recognised as significant threats to the quality of the Great Barrier Reef. Also since around 1981, however, very rapid expansion of the Queensland tourism industry has occurred, driven by increases in international tourism and domestic migration to Queensland. Tourist resorts were developed at Lizard, Green, Dunk, Magnetic and several of the Whitsunday Islands; coastal tourism facilities, such as the developments at Port Douglas and Port Hinchinbrook, have also expanded (Lawrence et al., 2002). In contrast, other Queensland industries have faced crises. The Queensland tobacco industry has ceased and declining sugar prices on the world market have reduced the profitability of sugar cane farming. Other industries – including commercial fisheries – are increasingly curtailed in the marine protected areas of the Queensland coast as they are believed to threaten the World Heritage status of the GBRWHA. Nevertheless, while the recent history of European activities in coastal Queensland is broadly characterised by rapid urbanisation and expansion of the tertiary sector, primary industries – mining, grazing, sugar cane farming and aquaculture – remain economically significant activities in the region (Powell, 1994; Bowen and Bowen, 2002).

Summary

This chapter has provided a brief overview of European settlement in Queensland, which followed a long period of natural and cultural changes since the formation of the Australasian continent. In summary, the period since European settlement in Queensland represents a time of considerable environmental change, although that period belongs within the larger history of the broader environmental impacts of European settlement in Australia and the environmental transformations wrought by Indigenous Australians (Archer et al., 1998). In Queensland, the first European settlers encountered an environment that had been extensively transformed by natural processes and by Indigenous Australian land management practices. In that context, the spread of European settlement, driven initially by pastoral expansion, was extremely rapid after the founding of Moreton Bay in 1824. By 1860, three major coastal ports were operating and, by 1920, most of the economically viable land in Queensland had been taken up by settlers. After 1860, sugar cane farming became the dominant form of agriculture in the colony and, later, extensive plantation production methods were replaced by the small cane farming system. By 1900, various environmental impacts had been sustained in coastal Queensland and the Great Barrier Reef, including the depletion of guano, *bêche-de-mer*, pearl oysters, dugongs, marine turtles, red cedar and rainforest areas, as the earliest terrestrial and marine industries became established. In the following decades, closer settlement and further land clearance also contributed to environmental degradation, which was compounded by the rapid expansion of tourism, commercial fisheries and mining during the second half of the twentieth century.

Significantly, since European settlement in Queensland commenced, many human impacts have been concentrated in coastal areas, increasing the interaction between land-use in the GBRCA and the condition of the nearshore waters of the Great Barrier Reef (Resource Assessment Commission, 1993). With its growing impact on coastal waters, European settlement in the region inevitably resulted in the degradation of the adjacent habitats of the Great Barrier Reef. The dominance of grazing, sugar cane farming and mining in Queensland means that those industries, in particular, are strongly implicated in the degradation of the Great Barrier Reef. Moreover, the very rapid spread of European influence in Queensland, the predominantly coastal European population and the relative economic specialisation of the colony means that environmental impacts have been concentrated geographically and temporally in coastal locations. Although the eastern coast of Queensland is long, the relative inaccessibility of large sections of it – particularly in Cape York – for most of the period of European settlement means that European impacts were localised, especially in the Cairns, Townsville, Whitsunday and Gladstone areas. Besides the indirect effects of terrestrial activities, considerable impacts on parts of the Great Barrier Reef also occurred directly as a result of many activities, such as guano and coral mining, coral and shell collecting, the introduction of exotic taxa (such as *Lantana spp.*) to islands, and dugong and marine turtle fishing. These and other impacts are discussed in the chapters that follow, which present evidence of some of the main changes in the coral reefs, islands and marine wildlife of the Great Barrier Reef.

Note

1 See the Annual Reports, Queensland Department of Mines, QPP, various years.

Chapter 5

The *bêche-de-mer*, pearl-shell and trochus fisheries

Introduction

Within a geomorphological context of significant vulnerability to degradation (outlined in Chapter 3), various human impacts on the Great Barrier Reef have occurred. One significant impact has occurred due to the operation of various fisheries based on reef resources, including *bêche-de-mer*, pearl-shell and trochus. Although reef organisms (including corals) have been removed from the Great Barrier Reef since the period of earliest European exploration, the first sustained European commercial fisheries in the Great Barrier Reef were the *bêche-de-mer* and pearl-shell fisheries. The animals that were harvested formed part of the landscape of coral reefs, and diving for those organisms was concentrated on, and in the vicinity of, those reefs. Large fishing grounds for each of those industries were located in Torres Strait, but the reefs of the Great Barrier Reef were also used extensively and, in some cases, the fisheries extended southwards as far as Moreton Bay. The earliest operation of those European reef fisheries was unregulated and few documentary records describe the beginning of those industries; Bauer (1964, p125) has acknowledged that production statistics for the *bêche-de-mer* fishery, for instance, were not available before 1884. The period of the historical *bêche-de-mer* and pearl-shell fisheries also lies beyond the range of oral history sources. However, the later development of those industries is described in Queensland Government records and reports – not least because of increasing concerns about the depletion of resources, and the abuse of Aboriginal and Torres Strait Islander workers – and the more recent trochus industry has also been described in oral history sources (Loos, 1982; Ganter, 1994; Reynolds, 2003).

The *bêche-de-mer* fishery

The early history of the *bêche-de-mer* fishery was first described in detail by Saville-Kent in his Annual Reports of the Queensland Chief Inspector of Fisheries that were published in the QVP and QPP (Saville-Kent, 1890a, 1893). The *bêche-de-mer* fishery began early in the European history of Queensland; one account attributes the earliest European commercial *bêche-de-mer* fishing to James Aicken at Wreck Reef in 1804 (MacKnight, 1976, p140). By 1827, bêche-de-mer were being exported from Cooktown and, by 1848, the remains of a bêche-de-mer smoke-house had been found by the crew of H.M.S. Rattlesnake. In 1857, J.S.V. Mein built a bêche-de-mer curing station at Green Island, which operated until the 1890s, and descriptions of that station were published in The Sydney Morning Herald (26 February 1866) and in the Cleveland Bay Express (19 April 1873). Many other curing stations were established in the Great Barrier Reef and, by 1880, bêche-de-mer stations were operating at Lizard Island, Green Island, Fitzroy Island, the Frankland Islands, the Barnard Islands and Dunk Island; in addition, the fishery at Cooktown employed thirteen vessels and two hundred workers (Serventy, 1955, p73; Jones, 1976, p16). Saville-Kent (1890a, p730; 1893, p231) reported that the period 1881–1883 was the most flourishing for the industry. By 1889, twenty-seven boats were operating from Cooktown, several boats each worked from Cairns, Ingham and Townsville, and a total of over 100 vessels were engaged in the trade.

The fishery was based on the collection of sea cucumbers (*Holothuria spp.*) from the substrate of the coral reefs. Saville-Kent (1890a, pp729–31) identified six commercial varieties of bêche-de-mer: teat-fish (*H. mammifera*), black-fish (*H. polymorpha*), red-fish (*H. rugosa*), prickly-fish or prickly-red (*H. hystrix*), lolly-fish (*H. vagabunda*) and sand-fish (*H. calcarea*). The names and values of those species in 1890 are shown in Table 5.1; the difficulties associated with the various scientific nomenclatures for bêche-de-mer species have been discussed by Skewes et al. (2004). Yet Saville-Kent acknowledged that scientific information about these species – including their breeding habits and growth rates – was scarce and he implied that a considerable lack of knowledge about the sustainability of the fishery existed. Nevertheless, a perception of plenty was articulated by some observers, such as Thorne (1876, p245), who stated that 'considerable quantities' of bêche-de-mer were found in the northern Great Barrier Reef, and Palmer (1879, p31), who wrote that Queensland's bêche-de-mer resource was 'extensive' and that

Table 5.1 Species and values of bêche-de-mer harvested in Queensland, 1890

Species	Local name	Value per ton
Holothuria mammifera	Teat-fish, black and ordinary	£140 to £150
	Teat-fish, white	£40
Holothuria rugosa	Red-fish, ordinary and deep water	£100 to £110
	Red-fish, surf	£80 to £90
Holothuria polymorpha	Black-fish, deep water	£110
	Black-fish, ordinary and Caledonian	£80 to £90
Holothuria vagabunda	Lolly-fish	£35
Holothuria hystrix	Prickly-fish (or prickly-red)	£30 to £40
Holothuria calcarea	Sand-fish	£20 to £30

Source: Based on data provided in Saville-Kent (1890a, pp730–1)

'thousands of tons of this valuable fish are to be obtained' by Sydney firms that were willing to invest in the industry. In 1879, Palmer (1879, p31) wrote that the revenue of the combined *bêche-de-mer* and pearl-shell fisheries was between £100,000 and £150,000 per year.

The geographical distribution of the *bêche-de-mer* fisheries extended as far south as the reefs to the east of Mackay, and as far north as Torres Strait; hence, the fisheries were concentrated in the northern Great Barrier Reef. The major centre of the fishery was located at Cooktown, with smaller centres at Cairns, Ingham and Townsville. In their bathymetrical distribution, most of the commercial varieties were found on coral reefs between 4 and 18 fathoms (approximately 7 to 33 metres) of water; the larger specimens of black-fish and red-fish were found at the deeper end of that range. The fishery used a system of small curing stations, at many locations in the Great Barrier Reef, from which small luggers – of 5 or 6 tons draught – made daily journeys to the reefs; alternatively, a fleet of luggers remained in the vicinity of the reefs, and used a tender to carry the catch to curing stations. In addition, a small number of schooners, weighing between 20 and 50 tons, were built at Cooktown and Thursday Island; those vessels carried portable curing facilities, as well as smaller boats and the processing equipment, and sometimes operated at sea for six months at a time (Saville-Kent, 1890a).

The average harvest for a *bêche-de-mer* station was around a ton of smoked product per month, as Saville-Kent (1890a, p729), describing the harvests and the collection methods, stated:

> A good average take for a fishing station working with only four boats, carrying twenty to twenty-four men, is one ton of cured *bêche-de-mer* per month. Two tons per month [...] represents an occasional but exceptionally abundant take. [...] The greater portion of the *bêche-de-mer* is simply picked off the reefs when the water has receded, but the finest red and black fish, and the prickly-fish almost exclusively, are obtained by diving during the same low tides from a depth of two or three fathoms.

The collecting process therefore involved *bêche-de-mer* fishers walking on the coral reefs at low tide as well as diving for the animals; some damage to corals must have occurred during the harvest. Saville-Kent (1890a, p729) stated that the animals were 'collected in sacks by wading or diving from off the reefs during the low spring tides'. In addition, some *bêche-de-mer* were caught as food for the boat crews.

Once the animals had been transported to the curing station, or to the schooner, they were smoked and dried. The process began when the fresh *bêche-de-mer* were placed in large iron cauldrons and boiled. The procedure, after boiling, was described by Saville-Kent (1890a, p729) in the following terms:

> The fish are then taken out, split up longitudinally with a sharp-pointed knife, gutted, and exposed on the ground in the sun until the greater portion

of the moisture has evaporated. The largest specimens, such as prickly and teat fish, are frequently spread open, so as to dry more readily, with small transversely-inserted wooden splints. The greater amount of moisture having been got rid of, the fish are transferred to the smoke-house. [...] The wood most in favour for the smoking process is that of the red mangrove, *Rhizophora mucronata*. Twenty-four hours is the usual period for which bêche-de-mer are left in the smoke-house [...].

After being smoked, the *bêche-de-mer* were bagged and transported to the nearest port, from which they were shipped to south-east Asia, particularly to markets in China (Suggate, 1940, p157).

The quantities and values of *bêche-de-mer* taken from the Great Barrier Reef during the period from 1880–1889 are shown in Figure 5.1. These graphs show the variable yields and returns obtained from the fishery; from 1881–1883, the fishery expanded, but fluctuations then characterised the *bêche-de-mer* fishery from 1884–1889, as Bauer (1964, p125) acknowledged. Yields declined in 1887; but, from 1887–1890, the fishery recovered and, by the latter date, over 100 boats were engaged in the trade (Saville-Kent 1890a, p730). Overall, the scale of the *bêche-de-mer* trade was substantial and of considerable importance to the colony; the returns from the trade made the *bêche-de-mer* fishery the second most profitable marine export from Queensland, after pearl-shell (Loos, 1982). However, the flourishing period of the fishery was short-lived. In 1890, Saville-Kent acknowledged the need for restrictions, surveillance of the fishery, and the appointment of an Inspector of Fisheries for the Cooktown district. In part, his concern derived from frequent reports of abuse of Aboriginal and Torres Strait Islander workers, and the fact that Indigenous workers were required to be registered at ports. The industry required little capital investment and paid low wages, generally to Aboriginal and Torres Strait Islander workers. Due to those difficulties, the profitability of the *bêche-de-mer* fishery declined during the 1890s and, in 1897, discussing the violence and inefficiency of the industry, Bennett (1897, p681) reported that the fishery was unsuccessful and that 'its total extinction would not be a matter for regret'.

Despite the worsening economic prospects for the industry for the period 1890–1900, some authors were optimistic about the wealth remaining in *bêche-de-mer* fishing. In 1899, Semon (1899, p246) wrote that the Great Barrier Reef 'is one of the richest *tripang* [sic] grounds existing, and it is continually ransacked by a lot of white fishermen from Thursday Island, Cooktown, and other north Australian settlements'. However, documentary evidence indicates that, by 1908, periodic, severe depletion of *bêche-de-mer* stocks had occurred. The 1908 Royal Commission investigation into the Queensland pearl-shell and *bêche-de-mer* industries collected oral history evidence from many fishers, who complained that little or no *bêche-de-mer* were available, and a closure of the fishery was recommended (Mackay et al., 1908, p. lxxiii). One *bêche-de-mer* fisher, Severin Berner Andreassen, reported that the animals had become scarce and few could

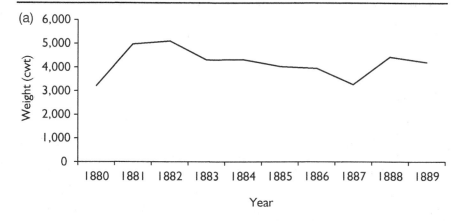

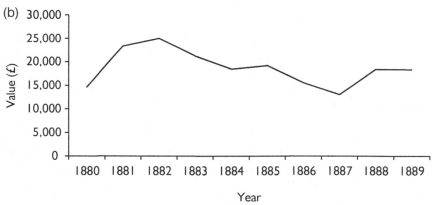

Figure 5.1 (a) Weights and (b) values of *bêche-de-mer* harvested in Queensland, 1880–1889. Source: Compiled from data provided in Saville-Kent (1890a, p730)

be harvested at Kennedy Reef, near Hinchinbrook Island. At all the places to the south of Cape Melville that he had visited, he claimed 'the reefs were skinned'; another area of particular exploitation was reported to be Endeavour Reef, where *bêche-de-mer* were scarcely available as a result of intensive harvesting in the Bloomfield River area (Mackay et al., 1908, p246). José Denis Antonio, a *bêche-de-mer* fisher at Bloomfield River, also reported severe depletion of the resources in that area, stating that as a result of continuous fishing, 'the reef has no chance' (Mackay et al., 1908, p240).

The report of the Royal Commission, written by Mackay et al. (1908), found that the animals – which were 'formerly plentiful' in the Great Barrier Reef – had 'either been exterminated there or driven to seek refuge in the deeper waters adjacent'; hence, divers were increasingly required to search for the animals in depths of 6 or 7 fathoms (around 12 metres) of water. As a result, those authors stated that the *bêche-de-mer* fishery seemed to have reached its

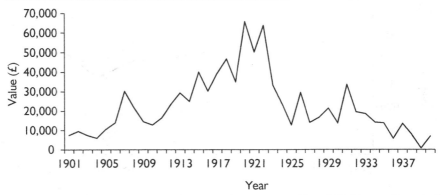

Figure 5.2 Values of *bêche-de-mer* harvested in Queensland, 1901–1940. Source: Compiled from data provided in NADC (1946, p44)

zenith in 1907 and that, since then, the reefs 'were fished bare'; consequently, they recommended a closure of the Queensland fishery for two years, enforced by a prohibition of *bêche-de-mer* exports from all Queensland ports. The Royal Commission concluded that the Queensland fishery was 'suffering from severe depression, which has resulted mainly from depletion of natural supplies' (Mackay et al., 1908, p. lxxiii). Although the complete closure of the fishery did not take place, a reduction in fishing effort was achieved by discontinuing the issue of licences for Asian vessels. The subsequent decline in revenue for the *bêche-de-mer* fishery is shown in Figure 5.2, which illustrates the fluctuating values of *bêche-de-mer* harvests in Queensland for the period 1901–1940, and the overall improvement in the profitability of the industry between 1910 and 1920.

In 1912, Mackellar (1912, p8) referred to the continuing operation of *bêche-de-mer* fishers in the Great Barrier Reef. By the end of the first decade of the twentieth century, an increase in the export value of *bêche-de-mer* had taken place, followed by a much greater expansion of the industry during and after the First World War, as Figure 5.2 suggests. In 1920 and 1922, the value of the harvests exceeded £60,000 in each year. Those years, however, represented the peak of the fishery and, after 1922, the fishery again declined. Some *bêche-de-mer* fishing continued during the 1930s; one lugger working near Green Island at that time is shown in Figure 5.3, and substantial quantities of *bêche-de-mer* continued to be removed from the reefs. For example, in 1933, the Townsville Harbour Board published its annual returns, stating that 86 tons 16 cwt of *bêche-de-mer* had been exported, and the trans-shipment of a further 11 tons 16 cwt 2 qtr had occurred. In September 1936, a cargo of 6 tons of *bêche-de-mer* was handled at Bowen Harbour by the A.U.S.N. Company. Another shipment, of 3 tons 19 cwt, was made in April 1937 (Anonymous, 1933a, p37).[1]

During the 1930s, some optimism about the fishery remained, as Glenne (1938, pp155–6) suggested:

Figure 5.3 A *bêche-de-mer* lugger near Green Island, c.1931. Source: Negative No. P05438, Cairns Historical Society Image Library, Cairns Historical Society, Cairns

There are no *trepang* fisheries like those of the Barrier Reef, where are found the *bêches-de-mer* [sic], which is neither a fish nor a slug, but an echinoderm. [...] Indifferently called sea-slugs, *trepangs*, *bêches-de-mer*, teat fish, or sea cucumbers, they are prized by the Chinese, who will pay as much £160 a ton for this beloved delicacy.

However, other authors acknowledged that the industry was declining. Suggate (1940, p157) stated that the 'quantities of *bêche-de-mer*, obtained, like tortoiseshell, from the coasts of Queensland and Northern Territory, seem to be declining', and he stated that *bêche-de-mer* collection took place in conjunction with pearl-shelling. By 1940, only small *bêche-de-mer* operations continued, as Figure 5.2 shows, including one fishery at Green Island. After that date, the fishery remained small although, in 1955, Serventy (1955, p76) stated that 'as much as £300 per ton was paid for this delicacy', and some *bêche-de-mer* fishing in the Great Barrier Reef has continued to the present day. Overall, the evidence presented above suggests that, by the time of the formation of the GBRMP, thousands of tons of *bêche-de-mer* had already been removed from the Great Barrier Reef. Recent scientific monitoring of the species has indicated that *bêche-de-mer* are now significantly depleted in the Great Barrier Reef as a result of the commercial fisheries (Uthicke and Benzies, 2000; Skewes et al., 2004; Uthicke, 2004; Uthicke et al., 2004).

The pearl-shell fishery

A detailed account of the Torres Strait pearl-shell industry has been provided by Ganter (1994); yet, although Torres Strait was the centre of that industry, reefs in the Great Barrier Reef were also exploited for pearl-shell (see Hedley, 1924). The Queensland pearl-shell fishery was the first to operate in Australia; the earliest pearl-shell raised in Queensland was taken from Warrior Reef, in Torres Strait, in 1868. The fishery sought *Meleagrina margaritifera*, the common mother-of-pearl, which occurred in two varieties of approximately equal abundance: the gold-lipped oyster and the common oyster. Of those two varieties, the common oyster, with a purer and more uniform nacrous lining, was the more valuable (Mackay et al., 1908, p. xlvi). A smaller pearl-shell, the black-lipped variety, was also found in Queensland waters as far south as Moreton Bay, but had not been harvested commercially by 1890. This variety was also known as 'Black Scotch', although uncertainty existed about its scientific name: M. *radiatus*, M. *fucatus* and M. *cummingii* were used variously to describe it (Saville-Kent, 1890a, p729). Pearl-shell became one of the most economically significant exports from Queensland, and was used in the manufacture of buttons and ornaments; the shell was exported to Europe and south-east Asia. While pearls were sometimes taken with the shells, those were not the commercial object of the trade and were usually kept by the divers (Saville-Kent, 1890a, p729; Glenne, 1938, p156).

The Queensland pearl-shell industry had its centre at Port Kennedy, on Thursday Island, where boats and crew members were registered and licensed, although pearling luggers worked shelling grounds in the Great Barrier Reef (Saville-Kent, 1890c, p704). Like the *bêche-de-mer* and trochus fisheries, pearl-shelling depended on Aboriginal and Torres Strait Islander divers before the introduction of diving equipment in the 1880s. However, the industry operated with little regulation and it was not until 1877 that production statistics became available for the Queensland pearl-shell fisheries (NADC, 1946, p8). In that year, Senior (1877, p311) wrote that pearl-shelling was 'a most thriving business' that had exported 200 tons of the material, at a value of around £200 per ton, from the port of Somerset in 1876. Senior (1877) stated that most of the luggers used – such as the vessel shown in Figure 5.4 – were owned by companies in Sydney; he wrote that no taxes or licence fees were required of those companies by the Queensland Government, and that a merchant in Birmingham had already purchased £30,000 of pearl-shell. Considerable optimism about the pearl-shell industry was expressed in Queensland. In 1879, Palmer (1879, p30) wrote that the coasts of Queensland 'abound in pearl-shell', and stated that near Cooktown he saw 'shells as large as dinner plates and about ¾ of an inch thick', worth from £150 to £190 per ton (see also Hedley, 1924, p5).

However, the depletion of pearl oysters had been recognised by 1897, when the Queensland Departmental Commission on Pearl-Shell and Bêche-de-Mer Fisheries was established to investigate the regulation of the fishery and to report the extent of exhaustion of pearl-shell resources (Hamilton et al., 1897, p1305). The following year, the Queensland Inspector of Pearl-Shell Fisheries, G.H. Bennett (1898, p1042), suggested that the whole of Endeavour Strait should be closed; that area, he stated, 'comprises grounds which have been constantly worked for many years, and from which large quantities of shell have been taken in the past, but it is now very much impoverished'. Bennett (1899, p995; 1900, p1319) acknowledged the need 'to close large areas of the pearling grounds for the purposes of conservation' so that the pearl oyster populations might recover, and he reiterated his concerns during the following two years, adding only that

Figure 5.4 The collection of pearl-shell aboard a Queensland lugger. Source: Negative No. 49810, Historical Photographs Collection, John Oxley Library, Brisbane

the need for the closure of the pearling grounds had become more urgent since there was practically no pearl-shell remaining in Endeavour Strait.

Yet in 1890, Saville-Kent (1890a, pp727–8) reported the continued profitability of the industry: he acknowledged that the pearl-shell fisheries of northern Queensland occupied 'a prominent position among the most important commercial industries of this Colony', and stated that, from 1884–1888, the average annual export value of pearl-shell was £69,000, more than double the combined value of the *bêche-de-mer* and oyster fisheries in Queensland. In 1890, Saville-Kent (1890a) reported, 1,000 workers were employed in the pearl-shell industry at Thursday Island, and 93 licences for pearling luggers were granted there, which was a reduction compared with the numbers operating before 1886. He attributed that decline to a large-scale migration of fishing operators to the Western Australian pearling grounds, although many of those operators subsequently returned to the Queensland fishery.

However, in spite of his comments about the profitability of the fishery, Saville-Kent recognised that some depletion of the pearl-shell beds had already occurred since it had become necessary to obtain pearl-shell from increasingly deep water as the shallow-water stocks became scarce. He stated:

> The average depth of water from which the greater quantity of the mother-of-pearl shell is at present collected is seven or eight fathoms [approximately fourteen metres]. In former years it was abundant, and is even now occasionally obtained in water of such little depth that it can be gathered with the hand at low spring tides. Twenty fathoms [approximately 37 metres] of water represent about the greatest depth from which the shell is profitably fished [...]. Some of the largest shell now placed on the market is collected at the above depth from off the New Guinea coast.
>
> (Saville-Kent, 1890a, pp727–8)

In his account, Saville-Kent reported that by 1890, as a result of the depletion of the earliest-harvested pearl-shell beds, the largest shells – weighing 8 lb per pair – that were once found commonly throughout Torres Strait had become scarce.

Saville-Kent acknowledged that the harvest of pearl-shell included very small pearl oysters. Describing the yields obtained by the fishery, he stated that a typical harvest was from 600 to 700 pairs of pearl shells per boat in one month: that represented approximately one ton in weight, although he acknowledged that, under very favourable conditions, 1,200–1,800 pairs of shells could be harvested, and that the owners of stations and boats awarded bonuses to divers and crews if they harvested over 1,000 pairs. One standard pair of shells was defined as 3 lb of pearl-shell and, although divers were encouraged to collect the largest shells, they were also able to obtain their bonuses by collecting very small pairs of shells if those amounted to the same weight; as a result, no incentive existed to preserve stocks of immature pearl-shell oysters. Consequently, Saville-Kent (1890a, pp727–8) reported 'a very considerable quantity of shell is brought in weighing from 1 lb to so little as 5 or 6 oz only per pair', which represented as many as 6,000 pairs of shells per ton. Furthermore, Saville-Kent acknowledged that the supply of pearl-shell was geographically variable, and that the most accessible beds had been depleted to a far greater extent than others.

Although Saville-Kent's report described the depletion occurring in Torres Strait, it is likely that similar depletion affected the northern Great Barrier Reef, since those pearling grounds were also used by the Queensland fishery. He argued that the depletion of the pearl-shell required immediate restriction and regulation of the industry. The decline in the average size of the pearl-shell harvested had reduced the value of the product; Saville-Kent (1890a, p729) stated that, because previously 'the price for shell of good quality ranged as high as £200 per ton, the shell itself was more readily accessible and [...] the profits in the trade were consequently much more considerable'. By 1890, however, the price of good quality shell had fallen to around £135 per ton. Saville-Kent (1890a, pp729, 734) reported that industry support for restrictions in the pearl-shell fishery had strengthened, and a trade body representing 73 boats had voted to accept a size limit of either seven inches from the front lip to the hinge overall, or of six inches across the diameter of the nacre; the latter measurement was preferable since the width of the surrounding border was highly variable, and that was the restriction that came into force.

These officials of the Queensland Government were not the only authorities to report on the decline of pearl-shell resources. In 1908, the Royal Commission investigation into the Queensland pearl-shell and *bêche-de-mer* fisheries acknowledged that, as old and full-grown pearl-shell had become scarce, the industry had adapted in an attempt to sustain yields: size limits had been imposed, pump-diving had been introduced, the average vessel size had increased and shore-station systems and pearling fleets had appeared (Mackay, 1908, pp. xlvi–xlvii). From 1890–1893, the statistics of the industry changed in the following ways: the number of boats increased from 92 to 210, the gross take of pearl-shell take increased from 632 to 1,214 tons, but the available catch per boat decreased from

6 tons 17cwt 1 qtr to 5 tons 15 cwt 2 qtr. Also, by 1893, a larger area was being fished for pearl-shell. Despite those changes, by 1894 the yield was stationary at 1,190 tons; by the following year the total harvest had fallen to 873 tons. By 1895, another source of pearl-shell had been found in Princess Charlotte Bay, but that resource was of inferior quality and may have contributed to a reduction in the market price for pearl-shell (Mackay, 1908, p. lxix).

The Royal Commission collected anecdotal evidence of the decline in pearl oysters using qualitative interviewing. The evidence suggested that the 'shallow beds inshore and those in the intermediate neighbourhood of the Prince of Wales group were the first to show signs of having been over-fished' (Mackay, 1908, pl). The causes of that depletion were thought to include the following eight reasons:

(i) ignorance about the length of time required for pearl-shell to mature;
(ii) a belief that the supply was inexhaustible;
(iii) the desire of pearl-shellers to raise as much shell as possible in the shortest space of time;
(iv) the introduction of floating stations, which concentrated the work of the vessels;
(v) excessive use of vessels;
(vi) the introduction of many Asian divers;
(vii) the lack of periodic closures of the fisheries; and
(viii) the reduction in size limits, from six to five inches (nacre measurement).

The Royal Commission concluded that the pearl-shell fishery was 'suffering from severe depression, which has resulted mainly from depletion of natural supplies'; consequently, urgent initiatives to cultivate the pearl oyster and to restrict the overseas labour force were required (Mackay, 1908, pp. li–lii, lxxv).

Another investigation into the industry – the 1913 Commonwealth of Australia Royal Commission on the Pearl-Shelling Industry (1913, p591) – found that the pearl-shell fishery was still 'capable, if systematically and scientifically conducted, of considerable development'. Individual pearl-shell divers were rewarded for large harvests using a system of incentives; the average annual harvest per diver was between six and seven tons, but divers were encouraged to take up to ten tons each year, and successful divers received a higher salary per ton. This system resulted in a large increase in the total pearl-shell yields during the periods 1911–1913 and 1918–1929; the yields obtained during the latter period were never exceeded in Queensland (NADC, 1946, p11). From 1912 to 1918, the value of pearl-shell had risen from £92,576 to £168,000, while the value of pearls during the same period increased from £25,000 to £63,000 (Taylor, 1925, p218). Nevertheless, after 1927, the industry declined as a result of the scarcity of pearl oysters. Between 1930 and 1934, the pearl-shell harvest decreased sharply and, subsequently, only a moderate improvement in yields occurred. In an attempt by the Commonwealth Government to support the struggling industry, a grant of £1,500 was made in 1935 to the Queensland fishery. In 1936, Christesen (1936, p31) wrote that the only remaining pearl-shell was found in deep-water beds, and Roughley (1936, p219) wrote that, although the resource was still available in Torres Strait, pearl-shell was smaller and less abundant to the south of Cairns.

The annual harvests and values of the Queensland pearl-shell fishery from 1890–1940 are shown in Figure 5.5, which illustrates the considerable variability that characterised the industry. In particular, the reductions in pearl-shell harvests during the First World War, and again during the early 1930s, are evident in Figure 5.5. These graphs show that, in a similar manner to the *bêche-de-mer* fishery, pearl-shelling reached its highest levels during the 1920s, although the peak of the latter industry occurred slightly later, in 1929, when 1,429 tons of pearl-shell were harvested. During the Second World War, pearling luggers were requisitioned by the Australian Navy, and no commercial pearl-shelling took place from 1941–1945. During those years, one report stated that:

> Exports of pearl shell was prohibited and later the Department of Munitions took over all the stocks in Australia, which were used for making prismatic compass dials for the Australian and Canadian armies, and to supply gold-lipped pearl shell for use as currency by the forces in New Guinea. Stocks fell so low that it became necessary to arrange for some pearl shell fishing.
> (NADC, 1946, p10)

Nevertheless, this small revival of the pearl-shell industry during the Second World War was short-lived as, subsequently, synthetic plastics replaced pearl-shell in the manufacture of buttons and the pearl-shell market collapsed.

A later report by the industry by the Northern Australia Development Committee (NADC), published in 1946, reached similar conclusions as the 1908 Queensland Royal Commission about the over-exploitation of resources. The report described the early phase of the industry, when pearl-shell was plentiful and could be collected from shallow water, and the necessity for divers to exploit increasingly deep stocks. The NADC acknowledged that until 1900 the Queensland fishery was far more successful than that of Western Australia, but that the Queensland fishery then declined, comparatively, until 1925. After that date, the Queensland industry recovered and dominated Australian production until its final collapse (NADC, 1946, p8). Yet the report acknowledged the severe over-exploitation of pearl oyster stocks, stating that:

> The beds had become very depleted, and of course the huge output of shell by the up-to-date Japanese ships had an adverse effect on the world market, and with the losses suffered by the Broome pearlers in the hurricane of 1935, left the Australian industry at a low ebb.
> (NADC, 1946, p9)

The NADC concluded that the Queensland pearl-shell fishery had operated on a basis that was far from sustainable and, hence, was comparatively short-lived.

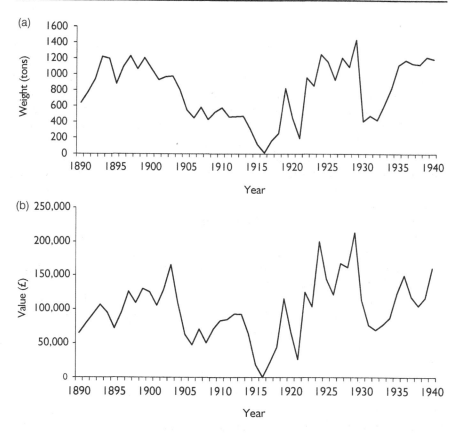

Figure 5.5 (a) Weights of pearl-shell harvested in Queensland, 1890–1940; (b) Values of pearl-shell harvested in Queensland, 1890–1940. Source: Compiled from data provided in Saville-Kent (1890a, p730; NADC, 1946, p44)

Serventy (1955, p75), writing in 1955, reported that both the value of pearl-shell and the cost of labour and transport in the industry fluctuated considerably; as a result, the high profitability of pearl-shelling before the First World War had decreased and manufacturers had turned increasingly to synthetic materials as substitutes for pearl-shell.[2] Like the *bêche-de-mer* industry, an accurate evaluation of the overall impact of the fishery on the resource is hindered by a lack of records for the early period of harvesting: few documentary records illuminate the period before the earliest depletion of pearl-shell was reported. Nevertheless, the evidence presented above suggests that the activities of pearl-shellers were widespread in the Great Barrier Reef, and that intense depletion of pearl-shell beds took place during the period in which that industry operated. Like the *bêche-de-mer* fishery, and in spite of regulatory measures, the pearl-shell industry exhausted the marine resources on which it was based.

The trochus fishery

The harvesting of trochus (*Trochus niloticus*) in the Great Barrier Reef occurred later than the main period of the *bêche-de-mer* and pearl-shell fisheries, since trochus was regarded as inferior to pearl oysters as a source of shell and was less sought after, although it was intended for the same purposes: button manufacture and the ornamental trade (Great Barrier Reef Fisheries Ltd, 1929, p15). Between at least 1912 and the 1950s, a large amount of trochus was harvested from the Great Barrier Reef. In 1912, Taylor (1925, p218) reported that the value of trochus in the Queensland fishery was £12,000, with most of the revenue derived from exports to Austria and Japan, and by 1916 the annual trochus harvest was around 500 tons. In July 1917, one shipment of trochus shell – weighing 5 tons 11 cwt – was handled in Bowen harbour by the A.U.S.N. Company; another shipment, of 6 tons 9 cwt, was transported in May 1936. Two further shipments of shell – weighing 3 tons 16 cwt, and 6 tons respectively – were handled at Bowen in July and August 1939.[3] The material continued to be collected primarily for export, as Suggate (1940, p157) reported, with the majority used to supply the Japanese market. The reefs of the northern Great Barrier Reef were extensively fished for trochus, and a large fishery also existed in Torres Strait, but the industry subsequently expanded to include the entire Great Barrier Reef, including the little-charted Swain Reefs in which trochus luggers operated as early as 1936 (Christesen, 1936, p28).

As the trochus fishery operated more recently than the *bêche-de-mer* and pearl-shell fisheries, some oral history evidence illuminates its operation. One former trochus diver described diving for trochus in the area between Cooktown and Sudbury Reef, near Cairns, stating that the crews worked the reefs 'till we had nineteen to twenty ton, [...] enough for the boat to carry. Used to call in Cooktown and put off some shells'. The same informant reported that the trochus crews worked for between six and eight months of the year, based at islands and harvesting many reefs between Cooktown and Cairns, and he stated: 'I was still on the boat at wartime. [...] The Americans used to buy the shell during the war, for making buttons'.[4] Another trochus diver recalled working many reefs between Cape York and Cooktown, stating:

> We seemed to work from a little below Somerset right down to almost Portland Roads. There were a lot of reefs and they'd work right along the reefs. And they'd usually anchor the boat on the leeward side and you'd row a dinghy and work your way back to the boat. So, depending on the tides, you were diving in shallow water up to quite deep: we went down about fourteen or fifteen feet.[5]

The same informant reported that trochus divers searched reef edges to find trochus; some carried the shells by hand, holding about a dozen at a time, but most used a small bag until the trochus could be emptied onto the boat. The processing of the animal took place on deck, as the same informant explained:

'they'd be boiled up and the meat dried away and it was dried out or something, because people used to eat it, and the shell would be bagged-up'.[6]

Another informant described the abundance of trochus on reefs in the Cairns area after the end of the Second World War.[7] In 1948, when he arrived in Cairns, that informant recalled seeing 'all the luggers' and the 'trochus shell just lying around' on the reefs.[8] That informant also recalled working on reefs near Mackay and as far south as the Swain Reefs; throughout the period of the fishery, he reported, trochus was plentiful on the reefs: 'it was everywhere'. He worked on board a ketch-rigged vessel, and stated that hundreds of luggers were working the reefs.[9] The same informant described the processing of trochus – and also the collection of an inferior type of shell, known as 'chicken-shell' – in the following terms:

> It fits inside a round tobacco tin; it's called chicken-shell and you're not really supposed to harvest it. They made a half-hearted effort to check luggers when they came in to see if there was any chicken-shell on board. The divers used to hide it. [...] Most of the boats fished one ton [of trochus] a day, with sixteen crew members on board, and they carried the old, square 44-gallon kerosene drums with a handle. [...] That's what they'd do the shelling in when the shell came on board. In the afternoon there would be a fire started in a 44-gallon drum that was split open a bit so that you could put another 44-gallon drum inside [...]. There was a hole cut out and you had mangrove water. That would make the water boil. Then you threw the shells in. After a few minutes you fished the shells out, put them on the deck, and all the crew would sit around with their piece of wire and their hook on the end: put that in the shell [...] and they'd pull the meat out.[10]

Another informant – also a former trochus diver – stated that in Cairns, in 1952, trochus-shell was valued at £500 per ton, and that divers were restricted to collecting shells no smaller than two inches in diameter in order to conserve trochus stocks.[11]

In comparison with pearl-shelling, trochus diving had the advantage of not requiring diving suits and mechanical breathing apparatus, because trochus could be found on the tops and the sides of reefs in shallow water, as one informant verified.[12] As Serventy (1955, pp75–6) stated:

> Instead of the boards used for the tenders which are a feature of pearling luggers, trochus boats have a large boiler attached to the stern. In here the trochus is boiled, the meat extracted and either eaten by the crew or smoked for sale ashore. The shell is packed ready for sale. It does not fetch the same high price as pearl shell and is also a much more fluctuating market.

As a result of the fluctuating market – and because of the size limits imposed – the trochus fishery did not result in such prolonged, widespread exploitation of coral reefs, in contrast to the *bêche-de-mer* and pearl-shell fisheries. The decline of the trochus industry was precipitated by the introduction of synthetic plastics in the 1950s, which caused the market for trochus to collapse, rather than by shortages of the natural material.[13]

Evidence of the survival of trochus stocks is found in several sources. In 1962, for example, Rees (1962, p102) stated that trochus shells up to eight inches in diameter could still be obtained from the Great Barrier Reef and were still used to manufacture buttons. In addition, monthly records of trochus taken from the Lizard Island group by the crew of the *Placid* indicate that, in February 1964, a total weight of 5 tons 5 cwt 1 qtr 11 lbs of trochus was taken, and in April 1964 the amount was 10 tons 11 cwt 2 qtr 10 lbs.[14] Since 1970, a limited market for trochus was re-established, and trochus fishing has continued in the Mackay area; the recent fishery, however, is small in comparison with the historical fishery that operated between 1912 and the 1950s. Overall, the evidence presented above suggests that the earlier fishery removed thousands of tons of trochus from reefs throughout the Great Barrier Reef; but, in contrast with the *bêche-de-mer* and pearl-shell industries, the operation of that fishery reduced – but did not exhaust – the supply of trochus.

Summary

This chapter has described one of the early European impacts on the Great Barrier Reef: the effects of the *bêche-de-mer*, pearl-shell and trochus fisheries. Strong evidence suggests that those industries caused sustained and intensive impacts on the marine resources they used. The earliest period of operation of the *bêche-de-mer* and pearl-shell fisheries, which dated from at least 1827, is not illuminated by production statistics, and was neither regulated nor monitored, with the result that the uncontrolled exploitation of resources took place. By 1890, Saville-Kent had reported the depletion of *bêche-de-mer* in Queensland; by 1898, the Queensland Inspector of Pearl-Shell Fisheries had acknowledged the scarcity of pearl oysters. In 1908, the Royal Commission into the *bêche-de-mer* and pearl-shell industries found that severe depletion of *bêche-de-mer* and pearl oysters had occurred, and that restrictions of the fisheries were necessary. Thousands of tons of *bêche-de-mer* were removed from the reefs, and early descriptions of the harvesting of these animals with ease from the reef flats at low water suggests an abundance that has not been described since 1922. Several contractions of those industries – each related to depletion of the marine resources – had occurred by 1950, after which date the fisheries declined to very low levels. By that time, coral reefs throughout the Great Barrier Reef had been harvested extensively by *bêche-de-mer* and trochus crews, and the reefs of the northern Great Barrier Reef had been exploited systematically for pearl-shell.

As a result of the extended period of operation of those fisheries, the *bêche-de-mer*, pearl-shell and trochus resources of the Great Barrier Reef were almost certainly significantly degraded from their status at the time of European contact by 1950. One scientific study of holothurian populations indicates that two species – the black teat-fish (*Holothuria whitmaei*) and the surf red-fish (*Actinopyga mauritania*) – are now regarded as overexploited; in addition, the sand-fish (*H. scabra*) has not yet recovered from fishing to very low levels (Skewes et al., 2004, p11). Skewes et al. (2004, p19) argued that holothurians are especially susceptible to exploitation because they are large and easily fished; they also stated that 'experience elsewhere has demonstrated that *bêche-de-mer* fisheries are extremely prone to overexploitation, and the recovery of depleted populations is slow and sporadic'. Skewes et al. (2004) acknowledged the particular depletion of the sandfish (*H. scabra*) as a result of overfishing in Torres Strait and in eastern coastal Queensland waters. However, they considered only recent (since the mid-1980s) fishing for *bêche-de-mer*. As Uthicke and Benzies (2000) and Uthicke (2004) have shown, in the Great Barrier Reef, the extent of the historical exploitation was very much larger than the recent fishery. Uthicke et al. (2004) have indicated the depleted status of black teatfish (*H. nobilis*) in the Great Barrier Reef; the authors acknowledged the very slow recovery of holothurian populations – which may take several decades – and advocated an extremely conservative management plan for that species. It is clear that, even decades after the most significant exploitation of these animals ceased, their populations have still far from recovered.

Notes

1 Harbour Board, Bowen, Statistical Book No. 3, January 1931–December 1945, RSI5551/1 Item 3, QSA.
2 The replacement of marine products with synthetic materials is also described in OHC 13, 4 August 2003.
3 Harbour Board, Bowen, Statistical Book No. 1, July 1915–February 1926, RSI5551/1 Item 1, QSA; Harbour Board, Bowen, Statistical Book No. 3, January 1931–December 1945, RSI5551/1 Item 3, QSA.
4 Fred Mundraby, cited in Thomson (1989, p85).
5 OHC 13, 4 August 2003.
6 OHC 13, 4 August 2003.
7 See Negative No. P09334, Cairns Historical Society Image Library, Cairns Historical Society, Cairns.
8 OHC 7, 19 February 2003.
9 OHC 7, 19 February 2003.
10 OHC 7, 19 February 2003.
11 OHC 22, 12 September 2003.
12 OHC 22, 12 September 2003.
13 OHC 7, 19 February 2003; see also Domm (1970, p45).
14 'Miscellaneous correspondence re. annual returns', RSI3284/1 Item 2, QSA; see also NADC (1946, p44).

Chapter 6

Impacts on marine turtles

Introduction

Of the seven extant species of marine turtle, six occur in Queensland waters: green (*Chelonia mydas*), hawksbill (*Eretmochelys imbricata*), loggerhead (*Caretta caretta*), flatback (*Natator depressus*), olive ridley (*Lepidochelys olivacea*) and leatherback (*Dermochelys coriacea*) turtles. Most are defined by the IUCN (2013) as critically endangered, endangered or vulnerable (Table 6.1). As a result of many natural and anthropogenic pressures – including predation of nests by feral foxes and pigs, incidental catches in fishing and shark control nets, ingestion of litter, boat strikes, Indigenous hunting, habitat destruction and tourism – declines in many marine turtle populations have been documented. Marine turtles are vulnerable to those impacts as a result of their life histories, which involve very high natural mortality of hatchlings and of small juvenile turtles, the use of a limited number of nesting beaches, high fidelity to nesting sites and feeding grounds, limited interaction between genetic stocks and long maturation periods. Mature female turtles come ashore at nesting sites on specific beaches to lay several large clutches of eggs in a single nesting season – a life history that requires high survivorship of adults. Yet many human impacts on turtles – including those described in this chapter – have affected female turtles disproportionately at that critical life stage.

Since European settlement in Queensland, various human activities have exploited marine turtles in the Great Barrier Reef and adjacent areas: the production of tortoise-shell, the commercial marine turtle fisheries, the 'sport' of turtle-riding at tourism resorts and the traditional hunting of turtles by Indigenous people. This chapter presents documentary and oral history evidence of the scale of those exploitative – and largely unsustainable – industries and activities. That evidence indicates that some intensive exploitation of those long-lived, slow-maturing animals – especially of green and hawksbill turtles – occurred. Consequently, observable declines in the numbers of those animals were reported. The account presented below indicates the scale of the exploitation that can occur in the absence of effective regulation or protection of vulnerable species and populations.

The commercial marine turtle fisheries in Queensland, which worked during the period 1867–1962, had a particularly severe impact on marine turtles because

Table 6.1 The IUCN Red List status of the marine turtle species of the GBRWHA

Common name	Scientific name	IUCN (2013) Red List status
Family: Chelonidae		
Loggerhead	Caretta caretta	Endangered
Green	Chelonia mydas	Endangered
Hawksbill	Eretmochelys imbricata	Critically Endangered
Flatback	Natator depressus	Data deficient (previously vulnerable)
Olive Ridley	Lepidochelys olivacea	Vulnerable
Family: Dermochelidae		
Leatherback	Dermochelys coriacea	Vulnerable

Source: Adapted from IUCN (2013)

they operated – at least in their earlier periods – with very limited, if any, government regulation or scientific monitoring. Given the lack of information about the early practices in that industry, precise reconstructions of the depletion of marine turtles cannot be made and estimates of former turtle population sizes, based on documentary and oral history sources, require careful interpretation based on current scientific knowledge of the ecology of those species. The actual magnitude of impacts on marine turtles is unlikely to be apparent for decades after those impacts occurred; moreover, the long-distance movements of individual turtles make changes in marine turtle populations difficult to monitor. Nevertheless, it is clear that, while the commercial fishery and other practices described below have ceased, human impacts on marine turtles in Queensland waters continue to require long-term, effective management, based on scientific research and monitoring linked with agreed environmental performance indicators (Limpus, 1997; Dobbs, 2001).

The tortoise-shell industry, 1871–1940s

The exploitation of at least two marine turtle species (green and hawksbill turtles) occurred in Queensland waters since the earliest European exploration in the region. The crews of vessels since the *Endeavour* caught green turtles as a source of fresh meat, and by the mid-nineteenth century the hawksbill turtle was recognised as a potentially valuable source of tortoise-shell. The tortoise-shell industry commenced in Queensland in 1871 when 20 lbs of tortoise-shell was exported to Great Britain; exports from that year until 1938 are illustrated in Figure 6.1, which shows that production was small in scale until around 1893. The tortoise-shell industry used the thick, overlapping scales that were taken from the carapace of the hawksbill turtle, which were an ideal export commodity as they could easily be dried and stored. Saville-Kent (1890a, 1893) reported that small quantities of tortoise-shell were also obtained from the green

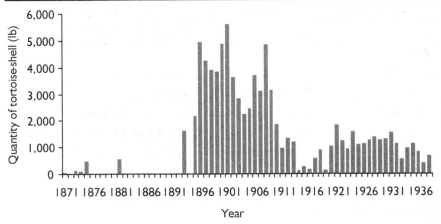

Figure 6.1 Exports of tortoise-shell from Queensland, 1871–1938. Source: Compiled from data provided in the Annual Reports, *QDHM*, 1895–1938, *QVP*, 1896–1900; *QPP*, 1901–1939; *SCQ*, 1871–1900; *SSQ*, 1901–1903

turtle, although those animals yielded a material of lower quality. Tortoise-shell production operated alongside *bêche-de-mer*, pearl-shell and oyster fishing, with the same crews and vessels sometimes being engaged in more than one fishery (Campbell, 1887; Bennett, 1898). Tortoise-shell production was concentrated in the northern Great Barrier Reef where the hawksbill turtle was abundant.

By 1889, Saville-Kent (1890a; 1893, p322) reported, the trade in tortoise-shell had increased: the average annual value of tortoise-shell exported from Queensland over the previous decade had exceeded £400, and its export value reached £1,705 in 1889. Saville-Kent (1893) stated that high-quality tortoise-shell reached a price of between £1 and £1 5s per imperial pound, although some highly sought variants could obtain a price of £20 per imperial pound. Very large quantities of tortoise-shell were exported from Queensland during the 1890s and 1900s, and rapid expansion of the industry had occurred by 1897, as shown by the increase in the number of tortoise-shelling vessels registered in Queensland between 1895 and 1897 (Figure 6.2). In 1898, the Queensland Inspector of Fisheries, G. H. Bennett (1898, p1048) stated: 'The supply of shell turtle seems to continue much the same, year by year, and affords an easy means of livelihood to the few coloured men [sic] engaged in it'. Two years later, Bennett (1900, p1319) reported that:

> [Tortoise-shell] is so valuable that shell-turtle is captured, when possible, by anyone who sees it, and has the means of attacking it – i.e., a boat and a spear. How the shell-turtle maintains its number in spite of all the enemies that pursue it – from the time the egg (an esteemed article of food) is laid on the sand beach through all the stages of its existence – is something of a mystery; but the fact remains that the shell-turtle appears to be as plentiful as ever, and its pursuit furnishes occupation and subsistence to a number of men [sic].

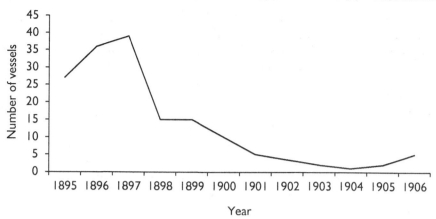

Figure 6.2 Numbers of tortoise-shelling vessels registered in Queensland, 1895–1906. Source: Compiled from data provided in the Annual Reports, QDHM, 1895–1938, QVP, 1896–1900; QPP, 1901–1939; SCQ, 1871–1900; SSQ, 1901–1903

His account suggests that, as early as 1900, the hawksbill turtle had experienced considerable exploitation, and that some doubts may already have been raised about the sustainability of the fishery. In 1901, the quantity of tortoise-shell exported from Queensland reached 5,579 lb (Figure 6.1); by the end of the following year, tortoise-shell had been shipped to New South Wales, Hong Kong, Ceylon, Germany and Great Britain.[1]

However, by 1908, concerns had been expressed about the extent of exploitation of hawksbill turtles, as evidence collected for the inquiry of the Royal Commission into the pearl-shell and *bêche-de-mer* industries reveals. Herbert Bowden, a pearl-sheller and merchant, reported that 'more notice should be taken of the present criminal action of men slaughtering turtle in the way they are doing'. He also stated: 'There is an enormous market for the turtle-shell itself. Hawksbill turtle is slaughtered wholesale for it' (Mackay et al., 1908, p197). Another pearl-sheller and merchant, Kenneth Ord Mackenzie, reported to the Royal Commission that the shell was removed from the backs of turtles using hot water while the animals were still alive, so that the shell could re-grow (Mackay et al., 1908, p129). However, the Royal Commission also heard evidence that the tortoise-shell industry had experienced a recent decline due to low prices for the product. Mackenzie reported that he had fished for tortoise-shell for a period of about six months and exported the material to London, but he stated that the catches were smaller and the market value was lower than he had anticipated, resulting in a small loss for his firm, Bowden and Mackenzie. Another merchant, Arthur Thomas Sullivan, also reported that the price of tortoise-shell was very low, although he argued that the industry remained profitable.

The declining profitability of the industry resulted in a contraction in fishing effort after 1897, a reduction in tortoise-shell vessels registered in Queensland from 1897–1904, and an associated decline in tortoise-shell exports from

Queensland after 1908 (Figures 6.1 and 6.2). However, changes in fishing practices may also have occurred in the industry after 1897, as the initial reduction in the number of vessels registered was not immediately associated with a decline in tortoise-shell exports. Instead, the industry may have become less specialised and hawksbill turtles were probably harvested in an opportunistic manner by the crews of vessels engaged in other fisheries, such as the *bêche-de-mer* fishery. New markets for tortoise-shell were found in Japan and the United States of America, although Great Britain remained the main destination for the product, and a small quantity of tortoise-shell was also exported to New South Wales.[2] Nonetheless, the price of tortoise-shell remained low and a sharp decline in the trade took place after 1908. Exports of tortoise-shell were particularly low during the years of the First World War (Figure 6.1). In 1916, J. R. Smith (1916, p1666), Inspector of Fisheries, stated: 'There is scarcely any demand for this shell at present'. Two years later, the Queensland Inspector of Pearl-Shell Fisheries, R. Holmes (1918, p1668), reported: 'The trade in connection with tortoiseshell has declined to such an extent that it is now a negligible quantity so far as the industry is concerned'.

A small revival of the tortoise-shell industry occurred after the First World War, which persisted until the outbreak of the Second World War. By June 1929, the value of tortoise-shell exports was £1,643, and Barrett (1943, pp40–1) wrote that in some years the annual revenue of the industry reached £2,000 or £3,000. Exports of tortoise-shell continued until at least 1938: production statistics published by the QDHM indicate that a total of 17 cwt of 'turtle-shell' was produced during 1933–1938 (Fison, 1935, p1104; 1936, p1155; 1937, p1414; 1938, p1295).[3] However, the trade never again resumed in Queensland on a scale comparable to that of the period before 1908. In the 1950s, synthetic materials replaced the use of tortoise-shell and the market for the natural product collapsed. Yet the evidence presented above indicates that by 1938 at least 86,020 lb (over 38 tons) of tortoise-shell had been exported from Queensland. As a result of the scale of this fishing effort, the tortoise-shell industry probably represented a significant impact on hawksbill turtle populations in northern Queensland waters, particularly since an incentive existed to harvest larger animals that yielded greater quantities of shell and that were more easily captured in the vicinity of traditional breeding sites, and also because of the low rates of growth and recruitment to adulthood that characterise the species.

Commercial marine turtle fishing in Queensland, 1867–1962

Besides tortoise-shell production, commercial fishing of marine turtles in Queensland occurred between 1867 and 1962 for the production of turtle meat and soup. Green turtles were exploited in the Great Barrier Reef and in Torres Strait; the emergence of two centres in the commercial fishery reflects the fact that two genetic stocks of green turtle exist, nesting in the Capricorn-Bunker

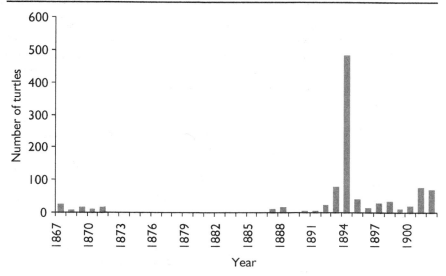

Figure 6.3 Exports of turtles from Queensland, 1867–1902. Source: Compiled from data provided in SCQ, 1870–1900; SSQ, 1901–1902

Group and at Bramble and Pandora Cays, respectively. The industry operated intermittently between 1867 and 1962, as substantial documentary and oral history evidence attests. However, the opportunistic exploitation of green turtles for food extended beyond those years. Early European explorers harvested turtles as a supplement to the ship's diet; the crew of the *Endeavour*, for example, took 21 large turtles in 27 days, near the Endeavour River, while the ship was being repaired (Limpus, 1980, p5). The harvesting of turtles in the Bunker Group commenced in 1803, during a voyage made by Ebenezer Bunker, in order to complement the provisions of sailing ships; that practice continued throughout the nineteenth century. In 1843, for example, H.M.S. *Fly* was stocked with turtles at Heron Island, and Jukes (1871, p172) stated that in the Bunker Group turtles 'in the greatest abundance were taken [...]. Turtle-soup, turtle-steaks, turtle-pie, and stewed flippers were our regular food for some time'. Around 1865, one report acknowledged the prevalence of turtles in Queensland bays (SPCK, c.1865, p254). However, the earliest evidence of the commercial harvest of turtles for food is found in the export statistics published in the SCQ, which indicate that the export of whole turtles from Queensland commenced in 1867; another report, of 1872, described a large harvest of 122 green turtles at Lake Creek, near Rockhampton, by one or two turtle-fishing boats.[4] Exports of turtles from Queensland for the period since 1867 are shown in Figure 6.3.

By 1886, a substantial turtle meat and soup industry had been established in the Moreton Bay area, where the animals were reported to be 'most plentiful in the summer months' (Fison, 1886, p835). A report by F. T. Campbell (1887, p123) stated that:

> [Green turtles] are caught in nets of a large mesh, and are mostly used in Brisbane at the hotels, and also preserved in tins as soup by Skinner of the Valley and other meat preservers. They are most common in the southern parts of the Bay, such as Russell Island, Swan Bay, and Broadwater, where a couple of men working industriously may take eight or ten per week. They are of large size and sometimes weigh 5 cwt.

In 1896, Saville-Kent indicated that preserved turtle meat was sought for the Chinese and other markets. Fison (1897, p637) stated that Peter Tuska had carried out a turtle-fishing operation in the Central-Moreton districts, where the supply of turtle meat was found to be 'sufficient for the moderate demand of the Brisbane Preserving Works'. Davitt (1898, pp261–2) provided the following account of turtles in the Fitzroy River estuary:

> On rising early in the morning to enjoy the view of the estuary of the Fitzroy, I was made aware of our having shipped several new passengers somewhere during the night. These were huge turtles, enormous monsters, so heavy that a sailor could not lift one of them. They are plentiful off the coast, inside the Barrier Reef, from Keppel Bay up northwards. They are, of course, cheap owing to their numbers, and it is customary, I was told, for drinking salons in the coastal towns to have turtle soup 'on tap' in the drinking bars [...].

By 1899, Fison (1899) reported, green turtles were still 'in fair supply when required' and, the following year, James H. Stevens (1900, p998), the next Inspector of Fisheries, stated that a 'good supply of turtle can nearly always be obtained when required'.

In 1900 and 1901, the turtle industry expanded in response to the introduction of refrigeration facilities, which facilitated the export of frozen turtle meat to London and Vancouver (Stevens, 1901, p1325). The Brisbane Fish Agency Company alone handled 70 green turtles during that year, obtaining £70 for those animals. Over the following two years, orders for turtles were easily met by that company – 53 animals being ordered in the first year – and 14,766 lbs of frozen turtle meat, in addition to 142 green turtles, were exported from Queensland. Moreover, the animals were reported to occur in large numbers in the Moreton Bay area, despite the increasing harvest (Stevens, 1902, p967). During the first decade of the twentieth century, more extensive exploitation of green turtles was prompted by the increasing demand in London for turtle products. In 1906, the Inspector of Fisheries acknowledged 'the valuable but as yet undeveloped trade in turtle' and the shipment of sun-dried turtle meat between Queensland and London (Stevens, 1906, p1419). Holmes (1933, cited in Limpus, 1978, p220) stated: 'One year several hundred [green] turtles were exported shell and all to the London market, and I believe they were a feature of the Lord Mayor's banquet'. During that decade, green turtles were also harvested from Masthead Island, in the Capricorn-Bunker Group (Figure 6.4).

Impacts on marine turtles 79

Figure 6.4 Loading a green turtle onto a boat using a winch, near Masthead Island, 1900s.
Source: Negative No. AP3:469, Robert Etheridge Collection, Photograph Archives, Australian Museum, Canberra

By the time of the Royal Commission enquiry into the pearl-shell and *bêche-de-mer* industries in 1908, commercial turtle meat and soup production was regarded as a promising industry. Bowden, the pearl-sheller and merchant, reported that turtle fishing could occur profitably alongside pearl-shelling, although insufficient knowledge of curing methods could cause the turtle meat to reach its destination in a poor condition (Mackay et al., 1908, p197). Nevertheless, Bowden reported that he had invested significantly in the industry. In addition, a merchant, Mackenzie, stated that he caught turtles and exported their shells, calipee (flippers), calipash (breast meat) and fat to London; the latter substance was used in soap manufacture. Mackenzie indicated that the problem of preserving the turtle meat had been overcome by keeping the animals alive on the decks of the ships. He reported that a turtle-tinning factory was operating at Rockhampton and that another works was run by Skinner in Brisbane, but he stated that the turtles he exported were obtained from Torres Strait rather than from the southern fishing grounds (Mackay et al., 1908, pp128–9).

More systematic turtle harvesting, based on nesting turtles, commenced in the Capricorn-Bunker Group in 1904 when Thomas Owens established a turtle factory on North West Island (Golding, 1979, pp48–9). His wife, Sarah Owens, wrote: 'To make one ton of extract it takes 440 turtles at 12 a day or 36 days. 100 cases of soup takes 228 turtles at 8 cases a day or 36 days' (Limpus, 1978, p221; Golding, 1979, pp50–1). Merchants in London reported that the demand for that turtle soup was unreliable, however, and in 1911 Stevens reported that there was little demand for turtles beyond the Sydney market, a situation that prevailed throughout the following decade, in spite of an apparent abundance of turtles in the waters of Queensland (Stevens, 1910, p927; 1911, p1192; 1913, p1033; 1920, p570). In 1919, Stevens (1919, p342) acknowledged that 'only a few' turtles were harvested due to insufficient demand. However, turtle-soup production at North West Island increased during the 1920s; by May 1924, the Barrier Reef Trading Company had constructed a canning works, wharf and rail track at North West Island (Golding, 1979). The scale of that operation was unprecedented in the Great Barrier Reef: during 1924–1925, 1,220 turtles were processed at North West Island and the annual output of the factory was approximately 36,000 tins (Forrester, 1925, p295; Barrett, 1930, p375).[5] In 1925, the Australian Turtle Company Limited commenced another operation at Heron Island, and in 1925–1926 the combined harvest of the two factories was 2,500 green turtles (Forrester, 1926, p929; Golding, 1979, pp61–2).[6]

The industry was intensive: a daily catch of about 25 female turtles yielded around 900 tins of turtle soup, with most of the harvest occurring in the nesting season from November to January. In addition to soup, tortoise-shell was sold, turtle shells and bones were used for the production of fertiliser, and turtle eggs were sold to biscuit manufacturers (Musgrave and Whitley, 1926; Napier, 1938, facing p136; QGTB, 1931; Roughley, 1936). In 1926, Forrester (1926) reported that regulation of the industry was necessary, and Musgrave and Whitley (1926) expressed concerns that the excessive harvest of nesting female green turtles at the two islands threatened the species with extinction. During 1926 and 1927, 2,475 turtles were taken, and 1,622 animals were caught the following year. Figure 6.5 indicates the large catches made between 1925 and 1928, when the equivalent of 136,000 twelve-ounce tins of turtle soup were produced at North West Island; at least 33,000 tins of soup were produced from 435 turtles during 1926–1929 at Heron Island (Forrester, 1927, p953; 1928, p1185; 1929, p953; Dick, 1930, p39). In addition to those catches, Barrett (1930, pp374–5) stated that turtles were found in abundance at Masthead Island, and green turtles were also taken from that island for the manufacture of turtle soup.

By 1929, green turtle catches in the Capricorn-Bunker Group had declined markedly. Several factors explain this contraction of the industry. First, as Limpus (1978) acknowledged, a shortage of reliable freshwater supplies on the cays hindered the boiling-down of the animals; other authors also acknowledged the problem of inadequate freshwater supplies, including the fact that *Pisonia* leaves contaminated the tanks at Heron Island, resulting in brackish water supplies (Musgrave and Whitley, 1926; Golding, 1979). Second, the quality of the tinned product was poor, as J. Huxham (1925, p15; 1928, p678), the Queensland

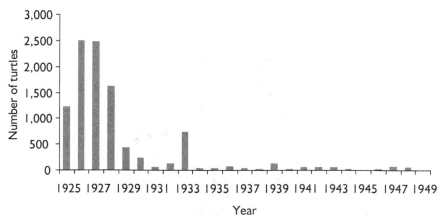

Figure 6.5 Numbers of green turtles harvested in the Capricorn-Bunker Group, 1925–1949. Source: Compiled from data provided in Annual Reports, QDHM, QPP, 1926–1950

Agent-General, reported: the twelve-ounce tins were too small, the consistency of the soup too thin, the content of green meat too low and the packaging too drab, he argued, for the London market. Third, the factory at North West Island was described as 'a somewhat ramshackle affair', and legal and financial difficulties hindered the operation of the Barrier Reef Trading Company. Fourth, as a report by F. W. Moorhouse (1935, p20) of the Great Barrier Reef Committee revealed, turtles had become scarce in the waters around Heron Island as a result of unsustainable fishing practices, including the capture of female turtles before they had laid their eggs. Fifth, and perhaps most importantly, as Limpus et al. (2003) have shown, large natural variations occur in the proportions of green turtle populations nesting in any one year. Consequently, the 1929–1930 turtle-canning operation was not completed. By 1932, the factories on both cays had closed and, during 1934 and 1935, the Heron Island factory was converted into a tourist resort.

On 15 December 1932, in response to calls for the protection of turtles at Heron Island, the Queensland Government prohibited turtle fishing during the months of October and November in waters to the south of latitude 17°S (to the north of that latitude, turtles were found to breed as late as May, so were not deemed to require legal protection). However, turtle fishing continued elsewhere and at other times of the year (Roughley, 1936, p255). Commercial turtle fishing continued in the vicinity of Gladstone to supply turtle soup and frozen turtle meat to the Brisbane and overseas markets. Green turtles were also captured at North West Island and transported alive via the Fitzroy River to the Lakes Creek Meatworks, in Rockhampton, to supply green turtle meat to the Central Queensland Meat Export Company Limited; that turtle-fishing operation is shown in Figure 6.6 (Dick, 1932). The market for turtle soup and frozen and sun-dried turtle meat was supplied by two companies: Great Barrier Reef Fisheries Ltd (1929) and Queensland Marine Industries Ltd (1932).

Figure 6.6 A turtle-fishing party on the Fitzroy River, c.1930. Source: Negative No. 13994, Historical Photographs Collection, John Oxley Library, Brisbane

Turtle fishing continued throughout the 1930s, but by the end of that decade the industry had declined in the Gladstone area and had become concentrated at the nesting site on Bramble Cay, in Torres Strait. During the years 1935–1936, Dick (1936, p1142) reported that 73 turtles were caught from the islands near Gladstone and in Torres Strait. In 1937–38, he stated that turtle fishing was confined to Torres Strait, where 30 animals were harvested (Dick, 1938, p1283). Production statistics published by the QDHM indicate that more than eighteen tons of shell-meat were produced during 1933–1938. During the 1940s, in contrast to the relatively small catches of turtles in the Capricorn-Bunker Group shown in Figure 6.5, more extensive catches were made in Torres Strait, bringing the total harvest of turtles, as reported by the Queensland Fish Board for the period 1938–1952, to 492 animals. However, catch rates were highly variable from one year to the next during that period, as the Annual Reports of the Queensland Fish Board demonstrate (Fison, 1935, p1104; 1936, p1155; 1937, p1414; 1938, p1295).[7]

Several Indigenous commercial turtle-fishing operations operated between 1940 and 1962 at Bramble Cay and in the Palm Island Group. One operation, carried out by the Genami Gia Turtle Fishing enterprise using a turtle trap in the Halifax area, was managed by the Palm Island Aboriginal Settlement.[8] A larger operation was carried out by the turtle fishing crew of the *Wanderlust*, based at Palm Island; their catch records between November 1940 and March 1941 indicate that 6,652 lbs of turtle meat without bone and 939 lbs of turtle flippers were produced (in addition to some dugong meat).[9] At Bramble Cay, the large-scale exploitation of green turtles was reported in 1953 by A. Mellor, Master of the *Melbidir*, who stated: 'It is nothing for each of the four boats to load 50 or 60 turtles during the night'. Mellor also reported seeing 'as many as ten full sized turtles dead on the Cay, where crews have turned them on their backs from previous raids, and departed loaded'.[10]

Other accounts of considerable destruction to turtles by commercial fishers were published in 1950. F. A. McNeill, the Curator in Invertebrates at the Australian Museum, wrote to the Honorary Secretary of the Great Barrier Reef Committee (GBRC) about the exploitation of green turtle numbers. McNeill stated that 16 to 18 green turtles were transported from Gladstone to Brisbane each week during the egg-laying season. He emphasised the disproportionate impacts on female green turtles, the inadequate recovery time between each 'period of butchering', the lack of enforcement of fishery regulations, and the 'conspicuous numerical dominance of males over females in the initial mating season'. Similar sentiments were expressed by the Queensland Government Tourist Bureau (QGTB), reflecting the concerns of many tourists about the destruction of green turtles and the cruelty of the methods used in the industry.[11] On 7 September 1950, in an attempt to halt the destruction of green turtles, an *Order in Council* prohibited the removal of green turtles or their eggs from Queensland waters and foreshores.[12]

Thus the commercial green turtle fisheries that had operated for almost five decades in the southern Great Barrier Reef ceased. However, Limpus et al. (2003) documented a brief revival in the industry that occurred in the northern Great Barrier Reef due to successful lobbying by commercial fishers; as a result, earlier legislation was repealed and a new *Order in Council* was passed on 4 September 1958 allowing the capture of green turtles in Queensland waters north of 15°S. Commercial fishing re-commenced and approximately 1,200 green turtles were caught in January and February of 1959 by the crew of the *Trader Horn*, although that harvest was not repeated due to doubts about the profitability of the enterprise. That final episode of commercial turtle fishing ended when all of Queensland's turtle species were designated as protected under the *Queensland Fisheries Act* by an *Order in Council* of 18 July 1968. The evidence presented here indicates that some intensive harvests of green turtles occurred in Queensland over the course of about a century, with thousands of animals – predominantly females – being caught for the production of turtle meat and soup. In 1999, the Queensland National Parks and Wildlife Service (QNPWS, 1999, p6) reported that the population of green turtles in the Capricorn-Bunker Group displayed 'some characteristics consistent with excessive loss of adult turtles from the population' – a loss that may be partly attributed to the impact of the commercial turtle fishery.

Figure 6.7 Turtle-riding at Masthead Island, 1900s. Source: Negative No. AP3:475, Robert Etheridge Photographs, Museum Archives, Australian Museum

Turtle-riding in the Capricorn-Bunker and Whitsunday Groups, 1900s–1960s

The recreational activity of turtle-riding involved the capture and use of turtles for sport at some of the tourist resorts of the Great Barrier Reef. Turtle-riding occurred in the Capricorn-Bunker group, which had large green turtle populations and which contained several tourist resorts; turtle-riding was particularly associated with the tourist resorts at Masthead, Heron and Lady Musgrave Islands. In addition to those locations, some turtle-riding took place in the Whitsunday Islands, including South Molle Island, and at Mon Repos Beach, Bundaberg. The origins of turtle-riding were attributed to Louis de Rougement, who first popularised accounts of that activity (Barrett, 1930, p375). However, turtle-riding was documented at Masthead Island as early as the 1900s, where the photograph reproduced in Figure 6.7 was taken by the palaeontologist and Director of the Australian Museum, Robert Etheridge Junior.

In 1926, when members of the Royal Zoological Society of New South Wales visited North West Island, Musgrave and Whitley (1926, p336) stated that, at that island, green turtles were put to another use besides the manufacture of turtle soup: 'By kneeling on their backs and clinging to the edges of the carapaces, we were able to ride them down to the beach and into the water'. Barrett (1930) reported that female turtles were captured at night as they came ashore to lay their eggs. The animals were overturned to prevent them from escaping and were left on the beach in that position until the following morning (Figure 6.8). At Heron Island, Barrett (1930, p360) stated, the 'turtle-strewn' beach and the spectacle of the animals, after their release, attempting to reach the sea 'provided the cameraman with an excellent shot'. The turtles

Impacts on marine turtles 85

Figure 6.8 Flatback turtles overturned on a resort island beach, 1900s. Source: Negative No. AP3:470, Robert Etheridge Photographs, Museum Archives, Australian Museum; Dr. C. L. Limpus, personal communication, 20 May 2013

were mounted by riders as they made their way down the beach to the sea (Figure 6.9). At Masthead Island, a 'Turtle Derby' was instigated, as Barrett (1930, p375) described:

> The united strength of three men [sic] was needed often to overturn a turtle caught napping on the sand. Grasping flippers and tail and the edge of the shield, we gave a sharp heave, and our friend was lying helpless on its back. The Turtle Derby of Masthead Islet was a memorable event! The jockeys wore bathing suits, and the 'winning post' was just the sunlit sea.

Figure 6.9 Turtle-riding at Heron Island, c.1930. Source: Photograph, Ref. 2849TB, QS57/1, Item 22, QSA

The establishment of the 'Turtle Derby' reflected the popularity of this sport, as several reports attest (QGTB, 1931; Anonymous, 1932; Holmes, 1933; Anonymous, 1937).

However, by 1940, concerns had been expressed that the use of turtles by tourists constituted cruelty to the animals. When an officer of the QGTB visited Lady Musgrave Island in 1940, he drew attention to 'the need for action to prevent the cruelty and destruction which some tourists are causing to turtles and other wildlife on the island'. He reported that the island contained thousands of turtle nests on the eastern, western and northern sides of the island, and he stated: 'Turtle life is so prolific on Lady Musgrave that Mrs Bell [the Caretaker] is taxed to the utmost in her endeavours to prevent cruelty to them by thoughtless tourists'.[13]

The impacts of turtle-riding on the animals – in particular, the habit of overturning the creatures during the preceding night – were considered in 1944 by A. M. Lewis, who had recently visited Heron Island and who wrote to the Queensland Society for the Prevention of Cruelty to Animals (QSPCA), drawing attention to the fact that the overturned turtles were often forgotten and left on the beach in blazing sunlight.[14] A. E. Cole, the Director of the QSPCA, wrote to the Secretary of the QDHM, stating that: 'I have personally seen turtles turned on their backs and left on the beach at Heron Island'.[15] In 1944, further complaints were made by visitors to Heron Island to the Director of the QGTB that turtles were mistreated and killed by the management of the tourist resort on the island.[16] Further evidence of cruelty and mortality to turtles in these locations was provided by Noel Monkman in a 1933 film entitled *Ocean Oddities*, and also by Glenne (1938).[17]

Turtle-riding did not cease with the introduction of legal measures to protect the green turtle in the Great Barrier Reef, in 1950; the activity was not included under the prohibition of the taking of green turtles, since the animals were not considered to be 'caught' as they were eventually allowed to return to their habitat. Turtle-riding in the Great Barrier Reef persisted until at least 1964, when turtles were used for that purpose at South Molle Island.[18] However, the activity has since been prohibited and all marine turtles now receive protection from human interference in the GBRWHA. The precise extent of the impacts of turtle-riding on the populations of the green turtle in the southern Great Barrier Reef, between the 1900s and the 1960s, is not known; however, the cessation of turtle-riding removed a source of interference with female green (and other) turtles during egg-laying, and probably reduced the availability of captive turtles that were vulnerable to other acts of cruelty or exploitation. The main period of turtle-riding coincided with the most intensive period of operation of the commercial green turtle fisheries, described above, and the associated decline in numbers of green turtles.

Turtle farming in Torres Strait, 1970–1979

The commercial harvest of marine turtles or their eggs in the Great Barrier Reef has been prohibited since 4 September 1962. However, the harvest of the animals and eggs by Aboriginal and Torres Strait Islander people formed an exception to that legislation, provided that the harvest took place for consumption only and without the use of explosives or poisons.[19] Attempts to farm turtles in Torres Straits were made during the 1970s, and large numbers of turtles and eggs were manipulated as a result of that activity. The turtle farming enterprise in Torres Straits was managed by Applied Ecology Pty Ltd, under the direction of Dr Robert Bustard of the Australian National University (ANU), with the support of the Queensland Department of Native Affairs (QDNA); it was hoped that the initiative would provide employment for Torres Strait Islander people who adopted turtle farming, in addition to providing a sustainable source of a culturally-important food item. Arrangements were made for the removal of juvenile turtles from Heron Island for the turtle farmers, since the juvenile animals were thought not to be available in sufficient numbers at Murray Island or Bramble Cay. As the proposed turtle farming did not fall under the exemption to the harvest of turtles permitted by Indigenous people for consumption only, a permit system was introduced, and arrangements were made for the Queensland Minister for Primary Industries to delegate authority to the Queensland Director of Aboriginal and Island Affairs to issue permits allowing Indigenous people who did not normally live on reserves to harvest sixty turtles (as well as thirty dugongs) per year.[20]

By December 1970, turtle farming had commenced in Torres Strait at Darnley and Murray Islands. Turtle pens, which were made of mangrove wood and which extended into the sea, were constructed to contain the animals. Green turtle eggs were imported to the farms; by September 1971, 2,000 green turtle eggs had been transported from Bountiful Island, near Mornington Island, to Darnley Island, where they were divided between farmers and reburied in the sand. However, the eggs did not hatch reliably after they had been moved; consequently, captive breeding of green turtles was attempted in Torres Strait. The intended markets for the turtle products were the Master Foods Corporation, which had manufactured turtle soup at a Sydney factory since around 1958, and which had received between 23 and 30 tons of produce, and the British soup maker and turtle merchant, John Lusty, who wished to launch turtle steaks on the London market. In addition to the farming of green turtles for food, Bustard proposed farming the hawksbill turtle for the manufacture of curios. He stated that 'the hawksbill turtle has been heavily over-exploited and is held to be rapidly reaching a position where it is directly threatened with worldwide extinction'. Bustard argued that the conservation of that species could be promoted by farming the animals in a sustainable manner and establishing a viable trade in hawksbill turtle products, including tortoise-shell and curios.[21] Other valuable turtle products included turtle oil ('which is used in large quantities by the cosmetics industry'), skins, leather and calipee.[22]

In April 1972, the farms at Darnley and Murray Islands were judged to have been successful and the industry expanded; new farms were established at Boigu, Yorke, Coconut, Yam, Stephen and Mornington Islands, and they were stocked with juvenile turtles collected from wild nests. On the reefs, the turtles were contained in permanent concrete pools measuring between 6 and 12 feet square. The collection of baby turtles proved to be difficult due to the natural variability of turtle egg-laying. Nonetheless, a large number of wild turtle eggs were collected and transported to the turtle farms, and wild turtles were also taken directly from the sea. One Murray Island turtle farmer, after collecting wild green turtle eggs in August 1971, obtained 181 healthy baby turtles for her farm. At Mornington Island, Bustard reported that one turtle farmer had caught many turtles (and dugongs), and that another farmer already had 25 green turtles.[23] Bustard described a turtle egg collecting trip as follows:

> On 13th March 1972 the Darnley Island turtle scheme chartered the *Ina* in order to visit Bramble Cay and collect this year's crop of hatchling turtles for the turtle farmers. We arranged to arrive at Bramble Cay in the early afternoon so that we could detect nests about to hatch and dig up the hatchlings. Baby turtles break out of their egg shells several days before they emerge on the sand surface. The baby turtles take up less space in the sand than the round eggs which do not pack closely together. This means that after hatching the sand in the neck of the nest falls down slightly so there is a small hollow. One can soon learn to detect these and dig straight down to the hatchlings which are making their way through the sand to the surface. In this way we secured over 500 baby turtles before darkness. [...] After dark we maintained constant patrols of all the beaches using pressure lamps and by 1.00am when the bulk of the hatchlings had emerged we had over 2000. [...] By the time we left shortly after dawn our total take of turtles had grown to slightly over 2500 baby green turtles.[24]

Therefore, the activities of the turtle farmers at Bramble Cay, during that season, resulted in considerable depletion of hatchlings.

Scientific knowledge developed alongside the commercial development of the turtle farms. At Long Island, for example, the first recorded nesting rookery for the hawksbill turtle in Australian waters was discovered. By 1972, the total number of turtles in the farms was between 5,000 and 6,000, with a target for the end of that year of between 12,000 and 15,000 turtles.[25] By April 1972, however, the first indications appeared that turtle farming was not taking place in conditions that promoted the survival of the animals, with outbreaks of fungal infection in captive turtles being reported. Since many turtles died in captivity, each farmer was required to hold more than 250 animals, and efforts to collect wild turtles to stock the farms increased. By May 1972, the problem of excessive mortality of turtles in captivity had become a significant issue for the turtle farming schemes.[26] Yet by that time the turtle farms had become well-established: both green and hawksbill turtles were farmed for the curios trade; a company had been formed – A. and I. Products Pty Ltd – to market the products; turtle farms had been established at Warraber, Yorke, Kubin and Mabuiag Islands; and turtle farming had expanded at Coconut Island.[27] On 8 February 1973, the Australian Prime Minister, Gough Whitlam, wrote: 'Some sixty turtle farms have now been established in North Queensland and three in Western Australia against a planned total of seventy-eight for the current financial year'.[28] In March 1973, between 2,000 and 3,000 juvenile green turtles were sought from Bramble Cay to supply the northernmost Western Islands in 1973.

Also in 1973, however, concerns about the conservation of the animals led to an attempt to restructure the industry to ensure the sustainability of the wild turtle populations that were being depleted in order to stock the farms. The Queensland Minister for Primary Industries wrote to the Queensland Minister for Conservation, Marine and Aboriginal Affairs, stating that the removal of 5,000 hawksbill turtles for the farms was acceptable, provided that at least 800 of the batch were released into the wild when they reached the age of one year.[29] By May 1973, over 20,000 turtles were being farmed on several of the Torres Strait Islands, and a new requirement was introduced: that at least 10 per cent of the farmed turtles should be released to the sea once they had attained a size 'adequate to ensure their safety from predators'.[30] Around that time, however, problems that had previously been overlooked in the industry became apparent. In particular, the statistics used to describe the scale of the industry were disputed; claims that 100 turtle farmers held 29,000 turtles were found to be inaccurate, as stocktaking found that around 19,000 turtles were held on the farms, suggesting that the mortality of farm turtles had been extremely high, partly due to the poor water quality and the cannibalism that was reported amongst turtles in overcrowded pens. Following debate about the industry, the Senate concluded that the schemes were inflicting excessive mortality on the animals – particularly on the hawksbill turtle, which had become 'almost extinct in Australia'.[31]

In November 1973, the turtle farming industry was completely reorganised in response to the criticisms of the Senators. A report by the House of Representatives Standing Committee on Environment and Conservation referred to the operation of 112 cottage industry farms in Torres Strait, farming around 29,000 turtles; but it acknowledged that a high mortality of turtle eggs and hatchlings had resulted from the activities of poorly-trained farmers. The Committee stated:

> The present system whereby Islanders establish themselves as turtle farmers by collecting sufficient eggs from rookeries to obtain 150 hatchling turtles appears to be causing harm to wild turtle populations. Some Islanders harvesting eggs from wild rookeries have reported a hatchling result as low as 4 from 700 eggs. Of those which do hatch only about 20% can be expected to survive.[32]

Particular failings of the industry included the fact that no records of turtle eggs removed – or of hatch rates – had been taken; turtle farmers were unsupervised and tended to exploit rookeries throughout the nesting season; and the rate of successful hatchling emergence in captivity was much lower than that found in wild turtle populations. In Torres Strait, captive turtles experienced a mortality rate of 80 per cent during the first month of their lives; larger turtles kept in overcrowded conditions succumbed to sickness and death from bloat.

The Committee found that previous efforts to ensure that 10 per cent of hatchlings were returned to the sea were ineffective: only eighteen turtles had been returned by mid-July 1973. Furthermore, those turtles that were returned were not healthy and undamaged; rather, blemished turtles that could not be sold as curios were returned and those experienced increased vulnerability to predation after their release. The Committee found that, in any case, given the excessive mortality rates that characterised the industry, a 10 per cent return rate of animals was far too low. In its evaluation of the overall impacts of the industry, the report stated: 'It appears to the Committee that the commercial aims of the enterprise have dwarfed the conservation aims'. Therefore, the Committee recommended: (a) that turtle farming should cease as a commercially-orientated undertaking; (b) that the activity should be re-established with an emphasis on research into the ecology of green and hawksbill turtles and on conservation *per se*, rather than on the exploitation of those species; and (c) that particular attention should be given to the conservation of the hawksbill turtle, which had become 'seriously depleted' throughout its range.[33] Subsequently, the industry was restructured to focus on research and monitoring activities. With the shift in emphasis from commercial to research activities, turtle farming declined in Torres Strait; in December 1978, E. Gibson wrote that the farms at Kubin, Coconut and Yam Islands had closed.[34]

Despite the reorganisation of the industry, concerns about the status of the turtle populations persisted. Limpus stated: 'Through most of its range the hawksbill turtle is considered a conservation problem'; as such, the species was considered to be endangered and actively threatened with extinction.[35] The turtle-farming operation – even if it was intended for purposes of scientific research – was by 1979 regarded as incompatible with the aims and methods of wildlife conservation. In September 1979, leaders of the Badu Island Council wrote to Charles Porter, the Queensland Minister for Aboriginal and Islanders Advancement, referring to the closure of the turtle farming operation. The Councillors were forced to consider alternatives to turtle farming, including fishing, although the latter activity was reported to be poor. The Council stated: 'Southern trawlers have been stripping these reefs of everything they can get over the past few years. They have cleared out many of the reefs'.[36] The cessation of turtle farming in Torres Strait left social and economic challenges to be faced by the former turtle farmers and their dependents; yet, with the passing of that industry, a source of disturbance to vulnerable turtle populations was removed.

Indigenous hunting of marine turtles

The hunting of marine turtles is a culturally significant activity for some Aboriginal and Torres Strait Islander people, who had developed considerable expertise in hunting turtles by the time of European settlement. McCarthy

(1955, p284) thought that Indigenous hunting of turtles probably amounted to many hundreds of animals per week. Saville-Kent (1890a, p733) acknowledged the skill of Torres Strait Islanders in capturing turtles. He described the use of 'sucking fish, *Echineis naucrates*' that were kept alive in the bottoms of canoes, fastened to pieces of line and released when a turtle was sighted; the fish secured themselves to the carapace of the turtle and could be used to haul in the catch. Additional details about Indigenous turtle hunting, including fishing methods and the use of turtle products, were provided by Haddon (1901). He reported that two main turtle harvesting seasons occurred: the first was called *surlangi* (in October and November), when mating turtles (*surlal*) could easily be speared at the surface; the second extended for the remainder of the year, when turtles were known as *waru* and were found in deeper waters and in channels between coral reefs. Haddon (1901, p155) also described the method of catching turtles using the sucker fish, which was known as *gapu* by the fishers.

An alternative method of capturing turtles, using spears, was reported by Wandandian (1912, p145). Sunter (1937, pp61–2) described the capture of turtles by Indigenous hunters using spears and 'throwing-sticks (*wommeras*)', but he indicated that the method of harpooning had also been adopted in order to catch turtles. Sunter (1937) stated that the turtles – if covered with wet bags – could survive for days after being harpooned, and reported that he had often carried the animals in that condition aboard his lugger. The traditional method of capturing turtles using the sucking fish, however, was not abandoned, as Glenne (1938, p155) reported; in addition, spearing of the animals still was practised, as Benham (1949) documented in the vicinity of Lindeman Island, and as Thomson (1956) described in Princess Charlotte Bay. The diverse methods of turtle hunting, including the use of dugout canoes and harpoons, were described in detail by McCarthy (1955, p284), who stated that the use of sucking fish to catch turtles was still practised by Aboriginal hunters between the Tully River and Cape York, although he claimed that the fish used was known as *Remora* rather than *Echineis naucrates*, and that the fish was used only to provide an indication of the movement of the turtle and not to haul in the animal (see also McGuire, 1939). McCarthy (1955, p284) acknowledged that Indigenous hunting of turtles occurred at Palm Island and along the northern coast from eastern Cape York to north-western Australia; therefore, he argued, the European harvest of the animals should be strictly regulated in order to conserve this important source of food for Indigenous people. McCarthy (1955, p283) stated that green, hawksbill and leatherback (or luth) turtles were the most vulnerable marine turtle species, and that the fishing of those species by Indigenous hunters had been 'one of the commonest sights in the old days, although not so much nowadays'.

The impact of Indigenous turtle hunting is difficult to assess from the documentary record; no evidence was found to indicate whether or not that harvest was ecologically sustainable. Mass (1975) noted that the animals were more readily caught during the mating season, and that females – usually containing eggs – were easier to capture than males, which may have had a disproportional effect on turtle populations. James (1962, p15) claimed that Torres Strait was once 'crowded with outriggers scouring the warm waters for turtles, but those days are gone'; yet the same author claimed:

> For hundreds, perhaps thousands of years, the [Torres Strait] islanders have hunted and killed turtles selectively, knowing that if they destroyed too many their food supply would diminish. Balanced killing indeed extended to all marine life in the [Torres] Strait.

In contrast to the actions of Indigenous turtle hunters, James (1962, p15) suggested that no concern for sustainability was shown by European settlers, 'who embarked on a systematic slaughter of the turtles [that] has continued practically to this day'. He acknowledged that the introduction of legislation by the Queensland Government was necessary in order to prevent the extermination of turtles in Queensland waters. However, that legislation was not always enforced; James (1962, p17) stated that 'a white fisherman boasted to me recently that he killed [turtles] whenever he saw them'.

Summary

This chapter has presented evidence of a range of human impacts on marine turtles in the Great Barrier Reef and its adjacent waters: the operation of the tortoise-shell industry and the commercial turtle fisheries; the recreational activity of turtle-riding at resort islands; turtle farming in Torres Strait; and Indigenous hunting of turtles. Those activities have had some severe impacts on turtles, especially for particular species, at particular times and in particular places. The hawksbill turtle, which was exploited heavily for tortoise-shell, is now listed as critically endangered by the IUCN. Green turtles were significantly depleted by the commercial fisheries – which impacted particularly on nesting adult females in the Capricorn-Bunker group, at a critical life stage – precipitating a collapse of the southern Great Barrier Reef green turtle fishery by 1932. Despite concerns about the lack of sustainability of commercial green turtle harvesting, the fishery continued in the southern Great Barrier Reef until 1950.

All of the historical impacts described in this chapter represent stresses that have combined to make marine turtle populations more susceptible to other anthropogenic influences, including boat strikes, by-catch in fishing nets and shark nets, marine pollution, Indigenous hunting and the disorientation of hatchlings by artificial lighting. Although the commercial exploitation of marine turtles – along with most of the other activities described in this chapter – has now ceased, the animals continue to require long-term scientific research and monitoring and effective management to reverse the declines in their populations. Moreover, the historical human impacts on turtles have increased the susceptibility of these animals to other (including natural) environmental changes, such as the degradation of important nesting habitat at Raine Island due to sand erosion, sea level rise and the flooding of turtle nests. At a time when marine turtle populations will need to be increasingly resilient to environmental changes, worldwide, it is clear that those of the Great Barrier Reef and its adjacent waters have barely begun to recover from their historical over-exploitation.

Notes

1. See the export statistics published in SCQ, 1875–1900; SSQ, 1901–1902.
2. See the export statistics published in SSQ, 1916–1924.
3. See the Annual Reports of the Queensland Fish Board for that period.
4. *The Rockhampton Bulletin*, 7 January 1872, cited in *Pugh's Almanac*, 1872; SCQ, 1870–1900; SSQ, 1901–1902.
5. See the Papers of Isobel Bennett, Box 6 Folder 10, Manuscript Collection, NLA.
6. See the Commissioner of Police to Under-Secretary, Queensland Home Department, Brisbane, 11 March 1925, Queensland Police Department, Commissioner's Office, Miscellaneous correspondence and reports, PRV10729/1 Box 169, Correspondence and reports – Inquiries re. turtles in Queensland, QSA.
7. See the Annual Reports of the Queensland Fish Board for that period.
8. SRS505/1 Box 520 Item 3625, QSA.
9. See the Reports of the Acting Superintendent, Palm Island Aboriginal Settlement, 'State of Receipts, Expenditure and Earnings of *Wanderlust* Turtle Fishing Crew, SRS505/1 Box 520 Item 3625, QSA.
10. A. Mellor, Master, QGPV 'Melbidir', Thursday Island to Mr C. O'Leary, Director of Native Affairs, Thursday Island, 22 December 1953, RSI5058/1 Item 1346, QSA.
11. F. A. McNeill, Curator in Invertebrates, Australian Museum, Sydney to Honorary Secretary, GBRC, Brisbane, 3 March 1950, RSI920/1 Item 9, General correspondence batches – General Tourist Bureau matters, QSA, pp1–4.
12. Director, QGTB, Brisbane to Under-Secretary, QDHM, Brisbane, 16 March 1950, RSI920/1 Item 9, QSA.
13. In-letter Ref. L40.2373.11, Secretary, Queensland Office of the Commissioner for Railways, Brisbane to Secretary, QDHM, Brisbane, 22 May 1940, PRV8340/1 Item 1, QSA.
14. A. M. Lewis to QSPCA, 22 January 1944, cited in A. E. Cole, Director, QSPCA to Secretary, QDHM, 28 January 1944, SRS5416/1 Box 10 Item 61, NP231, Bunker – Heron Island, QSA.
15. A. E. Cole, Director, QSPCA to Secretary, QDHM, 28 January 1944, SRS5416/1 Box 10 Item 61, NP231, Bunker – Heron Island, QSA.
16. Under-Secretary, Treasury, Brisbane to Secretary, Queensland Land Administration Board, Brisbane, 30 March 1944, SRS5416/1 Box 10 Item 61, NP231, Bunker – Heron Island, QSA.
17. N. Monkman, *Ocean Oddities*, Film recording, Australian Educational Films, Canberra, 1933, Title No. 18188, ScreenSound Australia, National Screen and Sound Archive, Canberra.
18. SRS189/1 Box 17 Item 73, Queensland Industry, Services, Views, People and Events; Photographic Proofs and Negatives; Islands – Barrier Reef, QSA.
19. *Order in Council*, 4 September 1962, RSI15058/1 Item 1386, General Correspondence, Marine Produce – Turtle Fishing – General Admin. Only File No. 1, QSA; 'Extract: Minutes Councillors' Conference, 1961', Ref. 9T/50, RSI15058/1 Item 1386, QSA.
20. In-letter Ref. 9T/50. Chairman, Badu Island Council to Deputy Director of Native Affairs, Thursday Island, 30 September 1962, RSI15058/1 Item 1386, QSA; Memo Ref. 23/11/1, Dr Bustard, Darnley Island, 11 January 1971, RSI15058/1 Item 1386,

QSA; Memo Ref. 37/13/1, Dr Bustard, Darnley Island, 13 January 1971, RSI15058/1 Item 1386, QSA; In-letter Ref. 71/7387. J. M. Harvey, Director-General, Queensland Department of Primary Industries (QDPI), Brisbane to Director, Queensland Department of Aboriginal and Islander Affairs (QDAIA), Brisbane, 29 March 1971, SRS505/1 Box 823 Item 5623, QSA.

21 *Torres Strait Turtle Farmers Newsletter*, no 1, June 1971, pp1–3, RSI15058/1 Item 1386, QSA; *Torres Strait Turtle Farmers Newsletter*, no 2, September 1971, pp7–12, RSI15058/1 Item 1386, QSA.

22 'Torres Strait Turtle Farming', c.1972, RSI15058/1 Item 1386, QSA; B. L. Venables, Agent for Nederveen and Co. Pty Ltd, Skin Buyer, Cairns to Manager, QDAIA, Brisbane, 10 January 1972, RSI15058/1 Item 1386, QSA.

23 *Torres Strait Turtle Farmers' Newsletter*, no 4, April 1972, pp2–4, RSI15058/1 Item 1386, QSA.

24 *Torres Strait Turtle Farmers' Newsletter*, no 4, April 1972, pp7–8, RSI15058/1 Item 1386, QSA.

25 *Torres Strait Turtle Farmers' Newsletter*, no 4, April 1972, p9, RSI15058/1 Item 1386, QSA.

26 Letters Ref. 9T/50/24/4, 101/26/4, Chairman, Coconut Island to Dr Bustard, Thursday Island, 26 April 1972; 9T/50 'A', 1/2/5, Chairman to Chairman, Stephen Island, 2 May 1972; 9T/50(A), 60/12/5, Dr Bustard, Thursday Island to [turtle farmers], 12 May 1972; In-letter Ref. 9T/50/18/5, JCM:PBT, Dr Bustard, Darnley Island to Director, QDAIA, 18 May 1972; In-letter Ref. 9T/50, Dr Bustard, Thursday Island to Director, QDAIA, Brisbane, 26 May 1972; 30/28/6; Dr H. R. Bustard, Thursday Island to Chairman, Sue Island, 28 June 1972, RSI15058/1 Item 1386, QSA.

27 Dr Bustard, Applied Ecology Pty Ltd, 'Torres Strait Turtle Farming Industry' (c.1973), pp2–3, RSI15058/1 Item 1387, General Correspondence, Marine Produce – Turtle Fishing – General Admin. Only File No. 2, QSA; Dr H. R. Bustard to Western Islands Representative, Badu Island, 24 February 1973, RSI5058/1 Item 1387, QSA; Letter Ref. 9T/50(A), 5/2/4, Dr Bustard, Darnley Island to Manager, Thursday Island, 2 April 1973, RSI5058/1 Item 1387, QSA.

28 In-letter, Hon. E. G. Whitlam, Prime Minister, Canberra to Hon. J. Bjelke-Petersen, Premier of Qld., Brisbane, 8 February 1973, RSI14900/1 Item 48, Ministerial Correspondence, Turtle Farming, QSA.

29 In-letter Ref. 9T/50(A), Queensland Minister for Primary Industries, QDPI, Brisbane to Hon. N. T. E. Hewitt, Queensland Minister for Conservation, Marine and Aboriginal Affairs, Brisbane, 18 April 1973, RSI5058/1 Item 1387, QSA.

30 In-letter, Chairman, Darnley Island to Hon. Mr Bjelke-Petersen, Premier of Queensland, Brisbane, 1 May 1973, RSI14900/1 Item 48, QSA; Out-letter, Director, QDAIA, Brisbane to Manager, QDAIA, Thursday Island, 9 August 1973, RSI5058/1 Item 1387, QSA.

31 'Adjournment: Aboriginal Affairs Ministry', 9 October 1973, Senate, pp1071–1081, RSI14900/1 Item 48 Turtle farming, QSA.

32 'Turtle farming in the Torres Strait Islands: Report from the House of Representatives Standing Committee on Environment and Conservation', November 1973, p17, RSI15058/1 Item 1387, QSA.

33 'Turtle farming in the Torres Strait Islands: Report from the House of Representatives Standing Committee on Environment and Conservation', November 1973, pp1, 6, 17–18, RSI15058/1 Item 1387, QSA.

34 E. Gibson, 'Turtle talk', December 1978, RSI15058/1 Item 1389, QSA.

35 C. J. Limpus, *Observations of the hawksbill turtle Eretmochelys imbricata (l.), nesting in north-eastern Australia*, QNPWS, Townsville, c.1979, RSI15058/1 Item 1389, QSA.

36 In-letter, Chairman, Deputy Chairman and Third Councillor, Badu Island Council, Badu to Hon. Charles Porter, Queensland Minister for Aboriginal and Islanders Advancement, Brisbane, 4 September 1979, p4, RSI15058/1 Item 1389, QSA.

Chapter 7

Impacts on dugongs

Introduction

The dugong (*Dugong dugon*) is a herbivorous marine mammal that occurs in, but is not restricted to, the GBRWHA, including the nearshore waters of the Great Barrier Reef which are considered to be one of the species' strongholds (Marsh et al., 2002, 2005, 2011). The dugong is listed as vulnerable to extinction due to various factors, including variations in seagrass availability, incidental drowning in shark nets set for bather protection, accidental by-catch in commercial gill nets, vessel strikes, habitat loss and over-fishing (IUCN, 2013). Since the animal is long-lived and slow-reproducing, with considerable investment in each dugong calf, the species is vulnerable to over-exploitation by humans. Consequently, the maintenance of dugong numbers, and the recovery of depleted dugong populations, is now difficult to achieve. Furthermore, dugongs are highly mobile animals that sometimes migrate across large distances, and in remote waters, with the result that impacts on dugongs are difficult to detect and assess (Marsh and Lawler, 2001; Marsh et al., 2004). Some previous assessments of the magnitude of European impacts on marine wildlife species, including dugongs, have been based on relatively uncritical use of secondary sources; those studies have been challenged recently on the basis of modern scientific methods and improved understanding of the ecology of marine mammals. Given the current conservation threats to dugongs, there is a need to use historical sources more critically to reconstruct the operation and likely impacts of the human activities that exploited those animals (Marsh et al., 2005).

The modern pressures on the species occur in the context of various forms of historical exploitation: those historical activities are the subject of this chapter. European commercial dugong fishing took place intermittently from 1847 until 1969 in the Moreton Bay area, and for shorter periods at other locations on the Queensland coast. Dugongs were also caught to supply dugong oil to Aboriginal settlements; in addition, dugongs have been hunted for their meat by Indigenous Australians – a practice which long pre-dates European settlement. The combined effect of those activities amounts to unsustainable exploitation – at least at particular times and in localised areas – that has probably reduced dugong numbers considerably. Declining dugong numbers were reported as early as the 1880s and the species required legal protection in 1888, indicating that dugong populations have been threatened by human activity for more than a century. Although the precise ecological impacts of those historical activities have not been established, it is likely that they – together with other impacts, such as variations in seagrass availability – have contributed to significant declines in dugong numbers.

Commercial dugong fishing in Queensland, 1847–1969

European commercial dugong fishing took place intermittently from 1847 until 1969 in the Moreton Bay area and for shorter periods at other locations on the Queensland coast. Dugong oil, hides, bones and meat were produced at stations in Moreton, Tin Can, Wide, Hervey and Rodds Bays, and also at a small dugong factory in Cardwell (Figures 7.1 and 7.2). Although the industry had declined by 1920, commercial dugong fishing occurred in the Moreton Bay area from 1847 until 1969 (Johnson, 2002). As with other fisheries in the Great Barrier Reef, commercial dugong fishing occurred with little regard for the sustainability of the harvest. Johnson (2002, pp27, 29) argued that such disregard resulted from a perception that the 'bounteous seas' would provide a limitless supply of animals. European commercial dugong fishing was pioneered by several dugong hunters at Amity Point, on North Stradbroke Island, Moreton Bay. The fishery was operating by January 1847, although Welsby (1907, pp52, 57) indicated that some dugong fishing occurred earlier at Amity Point, during the 1830s, and also at Pelican Banks. The fishery produced oil for cooking and for the production of cosmetics, which represented a new export from Queensland.[1]

The creation of a market for dugong oil is attributed to Dr William Hobbs, the Queensland Government Medical Officer, who encouraged the manufacture of medicinal dugong oil after 1852 as a substitute for cod liver oil. Hobbs exhibited dugong oil samples at the Sydney Museum and the 1855 Paris Exposition. His efforts prompted the establishment of a dugong fishing station at St Helena Island, in Moreton Bay, in 1856, and subsequently dugong fishing expanded northwards. From 1861, the availability and use of dugong oil was documented in *The Lancet* and *Pugh's Almanac* (1870). In addition to local demand, an order for 1,000 gallons of dugong oil was placed in 1862 by a British supplier of pharmacists (Thorne, 1876, pp257, 264; Loyau, 1897, p365; Welsby, 1907, pp86, 193).[2]

In 1860, Bennett (1860, pp165–6) described the methods used in the commercial dugong fishery, including the use of a floating station, harpoons and nets:

> A small cutter was fitted out early in the season, with a boiler for 'trying down' the oil [...] and the animal was to be harpooned in a manner similar to that by which whales are captured. The success, however, was so indifferent, that it did not pay the expenses, and was abandoned [...]. Since that time

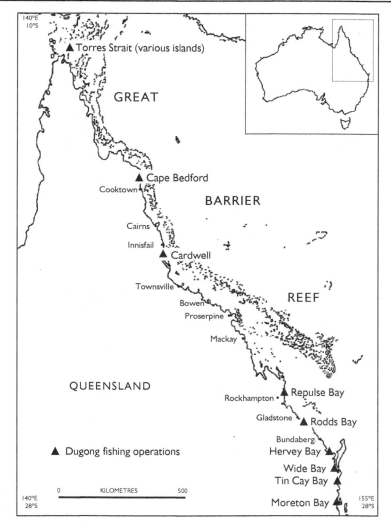

Figure 7.1 Locations of individual dugong fisheries in Queensland, 1840–1970. Source: Author

nets have been employed, and the result has been more productive. The nets are usually cast at night, in the places frequented by the animals, who become entangled in the meshes, and on average about two are captured every night.

The use of the floating station and harpoons suggests that the dugong fishery operated initially as a derivative of commercial whaling, although nets soon replaced harpoons. Bennett (1860, p166) documented the oil yields from large dugongs, stating: 'A full-grown animal yields from 10 to 12 gallons of oil. [...] Some

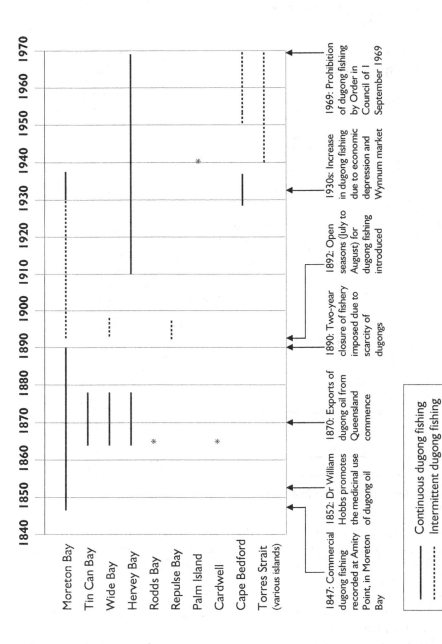

Figure 7.2 Periods of operation of individual dugong fisheries in Queensland, 1840–1970. Source: Author

are very large, and weigh from 8 cwt to half a ton', although he acknowledged wide variability in oil yields.

By 1864, in addition to the fishing stations in southern Queensland, a small dugong fishing station had commenced operating at Cardwell. By 1865, the possibility of a substantial dugong oil industry had been recognised as dugong numbers were reported to be very large in Queensland waters (Anonymous, 1861; Eden, 1872; Jones, 1961; Johnson, 2002). Describing what he regarded as 'a new and important branch of industry', Rowe (1865, pp123–4) stated:

> The dugong (*Halicore australis*) is abundant on all the eastern coasts of the colony [...]. Now that the oil is discovered to be valuable, it is exported to England in such quantities, that the fate of the dugong is sealed; and the fishery will eventually drive it in diminished numbers to the farthest and least approachable spots on the extreme north of the coast.

A report published in the *Brisbane Courier* in 1869, however, acknowledged that the distribution of dugongs in Queensland coastal waters was uneven: the animals were more numerous in Wide, Hervey and Rodds Bays than in Moreton Bay and were found 'at all seasons of the year in almost incredible numbers' in the tropical latitudes of Queensland (cited in Thorne, 1876, pp248–9).

From 1870, dugong oil, hides, tusks and bones were exported from Queensland to New South Wales, Victoria, Western Australia, Great Britain and Canada. Quantities of oil exported from Queensland from 1870 to 1902 are illustrated in Figure 7.3, which shows considerable fluctuations, although some large quantities of oil were shipped. In addition to the oil, 291 dugong hides and 4 cwt of dugong bones were exported to Great Britain between 1876 and 1878.[3] The hides were used to manufacture leather products; the bones were used to produce ornamental

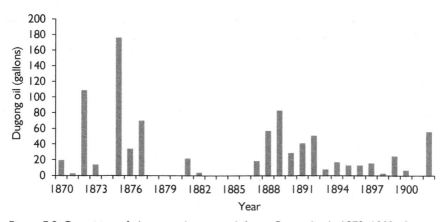

Figure 7.3 Quantities of dugong oil exported from Queensland, 1870–1902. Source: Compiled from data provided in SCQ, 1870–1900; SSQ, 1901–1902

cutlery handles. Soap was also manufactured using dugong stearin (the material that remained in the filters after the oil had been drained) and this product was exported to London. However, despite the lucrative prospects of the export trade, and Hobbs' efforts to secure an overseas market for dugong products, the majority of the dugong oil, hides and bones was sold at the Brisbane markets, while the meat was cured and sold – or given away – at the fishing stations (Johnson, 2002, p32).

In an account of 1876, Thorne (1876, pp248–9) acknowledged the abundance of dugongs and described the fishing operations in the Moreton Bay area, attributing their lack of success to the 'intemperance of the men employed'. During the 1870s, dugong processing stations continued to operate at Tin Can, Wide and Hervey Bays, although instances of contamination of dugong oil had by that time affected the market. Thorne (1876, pp254, 262) reported that an average dugong produced five or six gallons of oil and 100–200 pounds of meat, and that a large animal could weigh as much as a ton and could produce up to twenty gallons of oil: a much larger figure than Bennett (1860, p166) reported. Thorne (1876, pp260, 265–6) also described extremely large herds of thousands of dugongs, including one 'which appeared to fill the water' and was estimated to be 'half a mile wide and from three to four miles long'. Considering the apparent size of those herds and the extent of the dugong feeding grounds, Thorne saw no limit to the potential of the commercial dugong fishery.

During the 1880s, more systematic analyses of dugong numbers and behaviour were undertaken. Dugong meat was still readily sold in Brisbane, as Boyd (1882, p30) acknowledged, but an apparent scarcity of dugongs was reported by J. Lionel Ching of Maryborough, who stated: 'We left Great Sandy Island Strait in 1880, and since our return, eighteen months since, have found the dugong much scarcer' (cited in Fison, 1888, p764). In 1886, C. S. Fison, the Queensland Inspector of Fisheries, reported that observations of dugongs had been made along the coast; in 1887, he stated: 'There are a considerable number of [dugongs] still in Moreton Bay, and already one man here with only one net has been able to take 18 head during this season' (Fison, 1886, pp 833, 835; 1887, p123). Dugongs were sold for £5 each and were caught with long nets made from manila rope. A pattern to their abundance had also been noticed: dugongs were abundant in southern Queensland waters after flood seasons, but they could easily be driven away by boat traffic. The variable occurrence of the 'weed' (seagrass) on which the dugongs fed was also recognised (Welsby, 1907; 1931, pp57, 72).

Disputes had arisen about the organisation of the fishery. One concern was the advisability of expanding the existing system of dugong fishing licences. In 1888, Ching called for the wider use of licences, claiming that he was the only licensed dugong fisherman in Queensland, despite the fact that 'many parties' fished for the animal to the north of Wide Bay. A second issue was the acceptability of dugong fishing methods. Not all dugong fishers used the large-meshed nets that Ching had adopted; smaller-meshed nets prevented dugong calves from escaping, and the use of harpoons by other fishers led to the wasteful destruction of animals. Ching complained that the damage caused by harpooning was 'utterly ruining' his business, and he argued that the use of 36-inch mesh nets should be mandatory for dugong fishers. However, since a suckling dugong calf without a mother would be unlikely to survive, even the exclusive use of large mesh-size nets may have been an insufficient conservation measure. Ching also reported that he had imported a steam plant from England for boiling down dugong oil more efficiently (Fison, 1888, p764).

By 1888, the low reproductive rate of the dugong was beginning to be appreciated. Fison (1888, p764) stated that dugongs 'breed once a year, and have only one calf at a time'; modern research indicates that dugongs actually breed much less frequently than that (Marsh et al., 2011). Fison (1888, p764) also reported that, of the sixteen dugongs caught in Moreton Bay in 1888, almost all 'were cows in calf, or the young calves were found attending the mother'. The unsustainability of such a harvest was recognised, and restrictions on dugong fishing in Moreton Bay were introduced for a two-year period under the *Queensland Fisheries Act (1887)* in an attempt to prevent 'the utter extermination of the herd'. By 1889, the sporadic presence of a small herd of dugongs had been reported on the western side of Moreton Island, but Fison (1889, p939) stated that these animals were protected under the same clause of the *Queensland Fisheries Act (1887)*.

Fison's reports indicate that concerns about the unsustainability of the harvest prompted an early legal intervention to conserve dugongs in Queensland. The excessive destruction of dugongs was also reported by Saville-Kent (1890b, p713):

> A great amount of harm is done to the legitimate dugong fishery through the wasteful destruction of the animal in Wide Bay by means of harpooning, and also through the extensive slaughter of the young calves.

Saville-Kent (1890b) recommended that dugong fishing should be restricted exclusively to the use of stake-nets with a mesh-size of at least one yard square, which he believed would allow the calves to escape.

In 1892, the restrictions on the dugong fishery expired and commercial dugong fishing recommenced at North Stradbroke Island. Concerns about the unsustainability of the fishery had not been allayed and, in response to the scale of the resumption in dugong fishing, a second two-year closure was introduced on 1 January 1893. That closure was rescinded in June of the same year, however, after large herds of dugongs entered Moreton Bay; Johnson (2002, p36) associated that migration with extensive flooding in south-eastern Queensland during February 1892. The complete closure of the fishery was replaced with an annual three-month open season. Harpooning was prohibited and a more comprehensive licensing system was introduced (Welsby, 1907; 1931, p69). By the end of 1893, the Queensland commercial dugong fisheries had been regulated in similar ways to other commercial fisheries, with restrictions of permitted equipment and with spatial and temporal closures of fishing grounds.

In 1893, Saville-Kent (1893, p328) provided the following details of dugong processing at a station in Repulse Bay, near Mackay (see Figure 7.1):

> The hides, if well-cured, realise a price of 4½ d per lb, the large tusks of the male about half-a-crown per pair, while the bones make the best charcoal for sugar refining. [...] After many years' experience, it has been found at the Repulse Bay station that the old cows yield the most oil, the quantity being sometimes as much as eight or ten gallons, but on the average only four or five. The winter months, with respect to the amount of oil obtained, are the most profitable ones for the industry.

In contrast to southern Queensland waters, Saville-Kent (1893) stated that the northern Great Barrier Reef and Torres Strait had no systematic dugong fishery, although Indigenous hunting of dugongs occurred in those regions.

Fison (1894) reported that the dugong fishery in Moreton Bay was successful during the open season of 1894, yet by 1896 dugongs were again scarcely caught in Moreton Bay, although Fison (1897) later acknowledged that the animals remained numerous in the northern Great Barrier Reef. Thereafter the dugong industry failed to supply a significant market for oil and only a small, intermittent fishery existed between 1900 and the 1920s. In 1902, James H. Stevens (1902, p967), Queensland Inspector of Fisheries, stated:

> The demand for dugong during the last winter has not been a sufficient inducement for fishermen, although the fish [sic] has been plentiful. The only person who did start caught 30, and so did very well, selling the hides at 7 d per lb for carriage brakes. The flesh, too, served a good purpose [...].

In addition, 55 gallons of dugong oil were exported from Queensland to Victoria in 1902. Stevens attributed the lack of fishing effort to the collapse of the market that resulted from contamination of the dugong oil. Initially the impurity of the oil was blamed on negligence during the refining process; subsequently, samples of the oil were found to have been diluted with large quantities of shark oil. In 1905, Welsby (1905, p95) stated that in Moreton Bay dugongs were 'not nearly so numerous as in years past', and Stevens (1905, pp1041–2) reported that the dugong fishery in the Moreton Bay area was 'at a standstill' as only eight animals were caught during the season. Yet, he stated that high prices could still be obtained for dugong oil and hides if production were increased, and he acknowledged that much larger stocks of dugongs were found in coastal waters to the north of Wide Bay. The following year, using one net, the dugong catch in Moreton Bay increased to 45 dugongs, all of which were caught during an eight-week period (Stevens, 1906, p1419).[4]

By 1907, dugong fishing stations were still operating in Moreton Bay at Amity Point and Pelican Banks. However, the following year, the Moreton Bay fishery was again unsuccessful due to poor weather. By 1908, the oil was 'almost unobtainable'; a shipment of 70 bottles was sent to the Franco-British Exhibition that year, which represented almost all of the available supplies. In 1909, the scarcity of animals was attributed to a lack of seagrass at the Moreton and Boolong Banks, and no further dugong fishing was reported in 1910 (Stevens, 1908, p907; 1909, p931; 1910, p928; Welsby, 1907, pp52, 57). A small increase in activity occurred in Moreton Bay in 1911, with sixteen or seventeen large animals being caught at Boolong Banks; around that time, another dugong fishery commenced at Burrum Heads, in Hervey Bay, where dugong fishing was easy because 'the dugong were plentiful and caught very close to the beach' (Johnson, 1988, p77).[5] However, Stevens (1911, 1913, 1914) continued to express concerns about the decline of dugong numbers. His concerns were shared by Banfield (1913, p162), who reported that dugongs appeared to be less abundant in Hinchinbrook Channel than in previous years.

A small revival of the commercial dugong fishery took place during the First World War, prompted by shortages of cod-liver oil, but the revival of the fishery was short-lived because of the scarcity of dugongs. Limited dugong fishing continued during the 1920s. In 1923, the Amity Point station still supplied dugong oil, meat, hides, tusks and bones; by that year, dugong hides were being used to manufacture high-quality leather, engine belts and carriage brakes (Johnson, 2002, p37; Stevens, 1918, p1665; 1919, p342; Welsby, 1907, pp84–5). In 1923, however, a publication by the Queensland Government Intelligence and Tourist Bureau (QGITB, 1923, p23), compiled by the Great Barrier Reef Committee (GBRC), stated that dugongs were 'too rare to provide meat for the butchering trade', and expressed uncertainty about the survival of the species. In general, only intermittent fishing for dugongs took place in Moreton Bay during the 1920s, not only because of the reported scarcity of the animals but also because general fishing and oyster harvesting had become more lucrative (Stevens, 1920, p570; Dick, 1930, p40).

Determining dugong abundance was a complex issue. Concerns were expressed about declining dugong numbers; in 1913, for example, Banfield (1913) complained of the increasing rarity of dugongs in Hinchinbrook Channel due to their slaughter by Japanese trochus, *bêche-de-mer* and pearl-shell fishers. Nevertheless, dugong fishing continued at Burrum Heads, new operations commenced at Toogoom and in the Isis River area and anecdotal reports of large herds of dugongs continued to be made, including a sighting of at least 80 dugongs in Moreton Bay in 1928 (Allen, 1942, p533; Welsby, 1931, p56).[6] While some authors acknowledged that the vast herds once reported in Moreton Bay were no longer seen, uncertainty existed about the extent of the decline in dugong numbers: the animals may have become more timid or their herding behaviour may have altered. Welsby (1931, pp62–3) acknowledged that the distribution of dugongs corresponded to the distribution of seagrass, since dugongs were seen feeding even at the extreme edges of seagrass beds that uncovered at low tide. Dugong tracks could be observed through seagrass beds – even if the animals themselves could not be seen – and the ability of Indigenous fishers to 'read' the age of dugong tracks was exploited in setting dugong nets.

Yet the difficulty in estimating the size of dugong populations remained. Welsby (1931, pp103–4) stated: 'It is strange how many boating parties pass over the waters of Moreton Bay and never see a dugong', although he also stated that 'dozens and dozens' of dugongs were reported to be present in the waters between Wynnum and Amity Point and that the animals 'abound upon and around the Moreton, Amity and Pelican banks in hundreds the whole year through'. Such reports indicate the uncertainty that characterised early estimates of the dugong population in Moreton Bay. Yet that dugong population continued to support sporadic commercial fisheries, which probably resumed in response to the need for employment in the deteriorating economic conditions of the 1930s. Dugong fishing may also have been stimulated in Hervey Bay by the establishment of a company at Wynnum manufacturing a variety of fish and shark products (Dick, 1932, p6). One party fished for dugongs in Moreton Bay and in the Burrum River during 1930 and 1931, catching 35 animals of varying sizes in one month at the latter location; the following season, about 100 dugongs were caught in Hervey Bay. By 1935, dugong fishing was concentrated in Hervey Bay where nineteen dugongs were caught, followed by a larger harvest in 1936 in which one fisher caught 30 animals. In 1937 and 1938, 50 and 35 dugongs were caught, respectively (Dick, 1935, p1091; 1936, p1142; 1937, p1403; 1938, p1283; Lack, 1968, p5).

A summary of documented dugong catches in Moreton Bay between 1884 and 1938 is given in Figure 7.4. Uncertainty about the size of the dugong catches for the years 1912–1928, due to the omission of dugong harvest data from the Annual Reports of the Inspector of Fisheries, is apparent from the information missing from Figure 7.4; that omission probably reflects the comparatively low status of the dugong fishery in Queensland. However, Figure 7.4 illustrates the increased effort that took place in the fishery during the years 1930–1938.

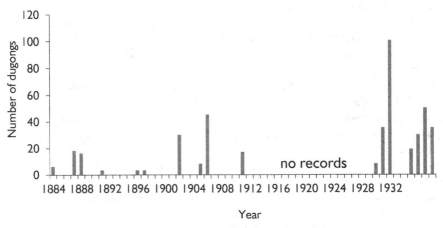

Figure 7.4 Numbers of dugongs caught in Moreton Bay, 1884–1938 (where records were available). Source: Compiled from data provided in *QVP*, 1884–1900; *QPP*, 1901–1938

Figure 7.5 Dugongs caught at Burrum Heads, c.1937. Source: Record No. 286431, State Library of Queensland, Brisbane

As a result, by around 1940, dugong populations were reported to be 'rapidly diminishing in numbers' (Tennant, c.1940, p49). Furthermore, no information was published about dugong fishing elsewhere. Lack (1968) acknowledged that commercial dugong fishing continued in 1940, with centres at Hervey Bay and Torres Strait, and that 70 dugongs were caught in Hervey Bay during that year. Intermittent fisheries continued to supply dugong oil for medical purposes and for the manufacture of cosmetic creams, with around 200 dugongs being caught annually at Burrum Heads alone: the annual harvest of that single operation (illustrated in Figure 7.5) represents approximately 12 per cent of the total estimated dugong population of Hervey Bay in 1999 (Preen and Marsh, 1995; Marsh and Lawler, 2001).[7] Given increasing concerns about the conservation of marine wildlife species, commercial dugong fishing was prohibited in 1969 by a Queensland *Order in Council*. By that year, Lack (1968, p5) stated, the dugong herds on the South Queensland coast had been almost wiped out, although scientific observations by Heinsohn et al. (1978) showed that large herds of dugongs were nonetheless found in Moreton Bay in 1978.

The supply of dugong oil to Indigenous settlements, 1928–1976

During the period between at least 1928 and 1976, dugong products – particularly dugong oil and meat – were supplied to Aboriginal and Torres Strait Islander communities in Queensland. In general, those products were obtained from fisheries that operated intermittently, and on a limited scale, in several areas. However, at some times and in some places, the capture of the animals occurred on a relatively intensive basis. The dugong oil was used for medicinal purposes and for cooking; it also represented a source of income for the missions. The dugongs were caught in the Starcke River and Palm Island areas and in Torres Strait, and the oil was sent to at least seven Indigenous settlements: Doomadgee, Lockhart River, Hope Vale (Cape Bedford), Yarrabah, Palm Island, Woorabinda and Cherbourg. In addition, dugong meat was also used to supply some of those settlements as well as other, local markets.

The report of the Chief Protector of Aborigines noted that in 1929 Hope Vale Mission augmented its food supplies by catching a large number of dugong, from which 180 gallons of dugong oil were sent to market. All of the dugongs caught for oil by Hope Vale were taken in the Starcke River area, about 70 km north of the mission. The men were taken up by boat to the camps near the hunting ground for about six months each year from about 1928 to 1936. The dugongs were caught using harpoons from small wooden dinghies. Estimates of numbers taken are difficult to determine: one informant reported that about 200 dugongs were taken per year for the three or four years he was involved. The dugongs were butchered, boiled down to extract the oil (using a Government-supplied extraction boiler) and the meat was then dried and sent back to the mission for food.[8] Smith (1987a) concluded that dugongs were captured for oil until at least 1936, then perhaps intermittently until the community was evacuated in 1942 during the Second World War. The intensity of hunting during that period apparently varied considerably, being inversely related to the effort put into the other mission-run marine industries (*bêche-de-mer* and trochus). The limited number of boats and experienced crew, as well as fluctuating market prices, largely determined which industry was exploited.

Archival evidences indicates that, during the 1940s, dugong oil was used for medicinal purposes (as an embrocation), as well as for cooking purposes, at Woorabinda Hospital, Cherbourg and Doomadgee Mission.[9] To supply the oil, small fishing operations commenced, based at Palm Island (using the *Wanderlust* turtle-fishing boat) and in Torres Strait. The dugongs caught were butchered and boiled down in steam-powered plants located at several stations, such as the plant at Cape Bedford (near Hope Vale). By July 1941, a processing station had been established at Badu Island, and a steam jacket plant for extracting dugong oil had been delivered to the island at the request of the Island Industries Board (IIB). By the end of 1941, dugong oil produced at Badu Island had been shipped to Cherbourg (160 gallons), Woorabinda (60 gallons), Palm Island (50 gallons), Yarrabah (48 gallons) and to Doomadgee (eight gallons); assuming that approximately four gallons of oil could be obtained from a large female dugong, those quantities of oil required the capture of more than 80 animals. By 1942, the supply of oil to the three settlements with the largest requirements had been organised into a system of standing orders placed with the IIB: Cherbourg received a regular delivery of sixteen gallons of oil per month, Palm Island eight gallons per month, and Woorabinda four gallons per month.[10]

By January 1942, oil supplies at Thursday Island were almost exhausted, although the Protector of Islanders predicted an increase in supply, as more crews would probably commence dugong-catching as the pearl-shell industry declined. The Second World War disrupted that predicted increase in dugong fishing; and, by 1944, the Queensland Director of Native Affairs wrote that the small amount of dugong oil being produced in Torres Strait was insufficient to meet local requirements. However, after 1945, production of dugong oil resumed; by 1949, supplies of oil had been obtained from Boigu and Saibai Islands, including one shipment of 136 gallons.[11] By 1950, another dugong processing station was operating at Saibai Island, where dugongs were captured using nets and the animals were either drowned or shot. The dugong oil from Saibai Island supplemented that of Boigu Island and was used to supply the mainland Aboriginal settlements. In 1951, however, the Director of Native Affairs reported that the only dugong oil produced in Torres Strait had been obtained from Boigu Island. By that year, a small trade in shark oil supplemented the dugong produce.[12] The Torres Strait stations not only supplied the mainland settlements with dugong oil, but also delivered dugong meat to local markets (Rohde, 1951).

Plans were also made in 1951 to revive the dugong oil industry at Hope Vale, after the community was re-established there after the Second World War.[13] In 1953, 255 litres of dugong oil (from at least fourteen dugongs) were collected and sold. Those plans came to an abrupt end in 1953 when the mission vessel sank. In 1955–1956, irregular trips were made to the hunting grounds. In one month in 1955, informants reported that approximately 150 dugongs were taken for oil and meat, but generally only irregular trips were made taking three or four dugongs per trip. No one interviewed at Hope Vale in the 1980s by Smith (1987a) could recall any hunting having occurred between 1956 and 1960. The early 1960s saw another attempt to revive the oil industry when a Hope Vale resident built an eleven-metre boat. As well as being used for other fishing activities, its three dinghies were used to hunt dugongs for oil and also meat (occasionally taken to the Mission by truck). In one month in 1961–1962, approximately 120 dugongs were taken for oil and meat. Outboard motors and rifles were introduced from the mid to late 1960s. Initially, rifles were used to kill the dugong after harpooning; however, occasionally dugongs were shot without harpooning. This method was unsatisfactory as the shot animals sank and were lost. Initially, outboard motors were used only to get to the hunting area; the boats were powered by oars for the actual hunt, but eventually motors were used throughout. Informants estimated that only about five dugongs per year were taken during this period, primarily for meat. In 1968–1969, requests were made for specific amounts of dugong oil to be supplied to a Cairns laboratory by the State Government, but those orders were not fulfilled and dugong hunting reverted to being a subsistence, ceremonial and recreational activity for the community (Smith, 1987a).

Archival evidence suggests that the production of oil in Torres Strait was variable during the two decades after 1951. In 1955, the Deputy Director of Native Affairs stated that dugong oil supplies at Thursday Island were exhausted, but that the oil could be obtained from Pastoral Supplies Ltd, in Brisbane. In 1958, Chemist Roush of Brisbane reported being unable to obtain supplies of dugong oil, although previously that retailer had purchased large quantities of the oil. Yet, by 1960, the drug store at Boigu Island reported receiving fresh stocks amounting to 140 gallons of dugong oil. The following year, two dugong fishers in Torres Strait reported that the animals had been 'scarce in their area', but they expected an improvement in their numbers during the north-west monsoon season, and an order for 100 gallons of dugong oil had been placed. By 1963, supplies of dugong oil had been replenished.[14] Fluctuations in the supply and demand of dugong oil continued during the 1960s. In 1966, the Deputy Director of Native Affairs wrote that dugong oil was urgently required at Thursday Island for medical purposes; by 1970, however, a surplus of dugong oil had accumulated. In contrast to the abundance of dugong oil in Torres Strait, the product was reported to be scarce in pharmacist shops in Cairns, although supplies of the oil could be obtained from Lockhart River Mission, which had for some time boiled its own stocks of dugong blubber.[15]

In 1969, the dugong was legally protected in Queensland waters, and by 1970 the commercial production of dugong oil for Indigenous communities had almost ceased, with the exception of small quantities of oil traded between several Torres Strait Islands. In 1970, the remaining stocks of 123 gallons of dugong oil at Boigu Island were sold. In place of the commercial dugong fishery, the introduction of a system of permits for Indigenous communities to harvest a limited numbers of dugongs and marine turtles was receiving consideration; by 1971, arrangements to allow Indigenous people who did not usually live on Reserves to take thirty dugongs and sixty turtles in any year were being debated. A small fishery for dugong oil continued in Torres Strait, and a small trade in dugong oil between those islands continued until at least 1976, with additional dugong fishing taking place at Mornington Island in the Gulf of Carpentaria. Nevertheless, the main period of dugong harvesting had ended.

Indigenous hunting of dugongs

In addition to the commercial fishing of dugongs described above, other harvests of dugongs have taken place since European settlement, including hunting of dugongs by Indigenous people (Holmes, 1933; Tilghman, 1933). Some details of the methods of dugong hunting by Indigenous communities on the east coast of the Cape York Peninsula have been provided by Smith (1987a, 1987b). Indigenous hunting of dugongs pre-dates European settlement in Queensland; that activity also continued throughout the period of European settlement and after 1969, when the commercial harvest of dugongs was prohibited. Indigenous hunting of dugongs is recorded in documentary sources since at least 1893, when Saville-Kent (1893) cited the ethnographic observations of dugong hunting in the western Torres Strait Islands made by Professor A. C. Haddon. Haddon (1901, pp148, 151–3) provided additional details of the methods used by Indigenous dugong hunters from Mabuiag to capture two animals during an earlier hunt, in October 1888, which exploited extensive fishing grounds at Orman's Reef

– between Mabuiag and the coast of New Guinea – and he also described the earlier practice of constructing a bamboo platform (*nēĕt*) on the reef from which dugongs were speared. Other documentary evidence of the Indigenous hunting of dugongs was provided by Banfield (1908), who acknowledged the apparent reduction of dugong numbers as a result of harvesting. In addition to the method of harpooning described by Haddon, Banfield (1908) referred to the practice of constructing and setting nets for dugongs.

A similar, early account of Indigenous hunting of dugongs, by Wandandian (Richard Dyatt) (1912), described the capture of a dugong cow in Trinity Bay, near Cairns, and the butchering of the animal to produce 528 lb of meat. Wandandian (1912) also referred to the spearing of seven dugongs in 90 minutes by one dugong hunter in Trinity Bay. Sunter (1936) provided an account of the hunting of a dugong by Indigenous people several years previously at Bowen Straits; he also acknowledged the apparent increasing scarcity of the animal in coastal Queensland waters (Sunter, 1937). The hunt took place using a canoe that was equipped with a spear and harpoon; the harpoon was thrust into its body and the boat was towed by the captive dugong. Eventually, exhausted, the dugong rose to the surface and was pulled into the canoe; the dugong was later butchered on the shore. Another method of Indigenous dugong hunting was described by Smart (1951, pp34–5), who described the use of a raft (*walpa*) of mangrove cedar by dugong hunters. Smart (1951) described the way in which hunters used the sound of the dugong exhaling to locate the animals; the dugongs were then speared when the raft was within striking range. In addition to the use of rafts, Smart (1951) described the method of fishing for dugongs using a system of wooden barriers and bark nets that were constructed across the entrances of small rivers. The dugongs were then driven into the nets using rafts. Once caught, the animals were drowned by holding them beneath the water. Smart (1951) also described an alternative method of dugong hunting and butchering used at Mornington Island, where the dugongs were harpooned from a canoe, towed in the water and drowned using a rope. Further details of Indigenous methods of capturing and butchering dugongs were provided by the anthropologist, Donald Fergusson Thomson (1956, pp33–6; 1985, pp14–15, 156–62), who referred to dugong hunting at several locations, including Princess Charlotte Bay, the Stewart River, Temple Bay, Cape Direction (Lockhart River) and Cape Sidmouth.

Whilst the Indigenous hunting of dugongs has exploited animals in Torres Strait and the Great Barrier Reef, that impact occurred alongside multiple other influences on dugongs, including the substantial harvests made by commercial dugong fishers prior to 1970. In addition, dugongs have been caught in shark nets set in Queensland for bather protection, which has resulted in the deaths of 654 dugongs in Queensland since 1962, and as a result of boat strikes (Marsh et al., 2002). Due to those other, anthropogenic factors, combined with natural environmental changes, dugong populations are now considerably influenced by harvesting by Indigenous hunters. Since the decline of commercial dugong fishing, the impact of Indigenous hunting now represents one of the most critical management issues for dugongs in Queensland, and the hunting of dugongs probably now occurs at unsustainable rates in Torres Strait and the northern GBRWHA (Marsh et al., 2004, 2005, 2011).

Summary

This chapter has described some of the historical impacts on dugongs that have occurred in Queensland waters. Commercial dugong fishing in the Moreton Bay area resulted in the reported local scarcity of the animals by 1888; by that year, the methods of catching dugongs were considered to have caused excessive destruction of the animals and restrictions had been introduced. In 1923, concerns about the survival of the species were reiterated. Nonetheless, commercial dugong fishing expanded in the 1930s and persisted until 1969, when the dugong received legal protection from commercial fishing; during the latter period, however, a single dugong fishing operation in Hervey Bay was responsible for a reported annual harvest of about 200 animals. Dugong populations were depleted by such commercial operations for more than eighty years after concerns for their numbers had first been expressed, and the surviving dugong population in the Toogoom-Burrum Heads Bay in 2003 was estimated by one dugong fisher to be about 200–300 animals: approximately the size of the annual harvest of the dugong fishery at Burrum Heads. However, in 1999, the dugong population of Hervey Bay was estimated by Marsh and Lawler (2001, p125) to be approximately 1,650 animals.

In addition to the impacts of the commercial fishery, dugong numbers were also reduced in order to supply dugong oil to Indigenous settlements, between 1928 and 1976. That activity apparently took fewer dugongs from the more robust northern Great Barrier Reef population and produced smaller quantities of oil than the commercial industries, servicing local communities rather than the larger markets that were supplied by the commercial fisheries; nonetheless, that smaller fishery represented another layer of impact on the same stock that had already been depleted in the southern Great Barrier Reef. The effects of both fisheries were compounded by Indigenous hunting of dugongs, which still continues and is of cultural as well as economic significance to Aboriginal and Torres Strait Islander people. The degradation of some seagrass beds as a result of terrestrial run-off associated with extreme climatic events, the prevalence of accidental dugong mortality due to boat strikes, and the drowning of dugongs in shark control nets set for bather protection since the 1960s now form serious threats to the species. Furthermore, Indigenous hunting, which now represents the largest single impact on dugong populations in the GBRWHA, is largely unchecked (Preen and Marsh, 1995; Heinsohn et al., 2004). Any additional activity that increases the mortality of this vulnerable, long-lived and slow-reproducing species now conflicts with the World Heritage values of the ecosystem.

Since the formation of the GBRMP, scientific knowledge of dugongs has greatly improved, however, placing the narrative presented above in a broader context. Heinsohn et al. (1978) indicated that a resident population of at least 300 dugongs remained in Moreton Bay in 1978; those authors also suggested that significant migrations of dugongs between feeding areas occur, although that suggestion had not been scientifically verified at that time. A more recent study, by Chilvers et al. (2005), described the existence of large populations of marine mammals, including dugongs, in Moreton Bay, adjacent to highly-developed coastal environments, and despite the historical exploitation of the animals. Those studies suggest that caution is required in reconstructing the impacts of the commercial dugong fisheries, since the effects of over-exploitation may have been confounded by those of large scale movements of the animals. That confounding would account for the scarcity of dugongs reported in the historical literature at some locations, in some years, and the biologically-impossible apparent rapid recovery of the population. In particular, the fluctuations in dugong numbers observed in Moreton Bay may be attributed to dugongs moving in response to changes in their food supplies, as Marsh et al. (2004, 2005) suggested, as well as a response to exploitation. Reconstructions of historical marine wildlife species populations are problematic, as Marsh et al. (2005) have shown, and the response of marine animals – such as dugongs – to anthropogenic and natural pressures may be complex; therefore, further scientific research and monitoring is required to understand changes in dugong abundance.

Notes

1. *Sydney Morning Herald*, 25 January 1847, p2.
2. *Brisbane Courier*, 4 September 1862, p2; *Brisbane Courier*, 31 July 1863; *Pugh's Almanac*, 1861, p48.
3. These statistics were compiled from SCQ (for the years 1870–1900) and SSQ (for the years 1901–02).
4. See the statistics published in SSQ, 1902, p199.
5. OHC 34, 12 October 2003.
6. OHC 34, 12 October 2003.
7. OHC 34, 12 October 2003.
8. See the annual report by J. K. Bleakley, Director of the Queensland Department of Native Affairs (QDNA, for 1929, published in QVP.
9. Superintendent, Woorabinda Aboriginal Settlement to Director of Native Affairs, Brisbane, 30 December 1940, SRS505/1 Box 662 Item 4493, Correspondence Files (Alphanumeric), Woorabinda – Medical – Supplies, Dugong Oil, QSA; C. G. Brown, Superintendent, Yarrabah Mission, Cairns to Director of Native Affairs, Brisbane, 21 February 1941, SRS505/1 Box 1028 Item 7033, Correspondence Files (Alphanumeric), Administration – Yarrabah – Supplies, Dugong Oil, QSA; Deputy Director of Native Affairs to Acting Superintendent, Cherbourg Aboriginal Settlement, 12 March 1941, SRS505/1 Box 585 Item 4027, Correspondence Files (Alphanumeric), Cherbourg – Medical – Supplies, Dugong Oil, QSA; Matron Peatry, Woorabinda Hospital, 13 September 1941, SRS505/1 Box 662 Item 4493, QSA.
10. Acting Superintendent, Palm Island Aboriginal Settlement, 9 January 1941, 'State of Receipts, Expenditure and Earnings of "Wanderlust" Turtle Fishing Crew for month ended 31/12/40 including Trip ended 22/11/40', SRS505/1 Box 520 Item 3625, Correspondence Files (Alphanumeric), Palm Island – Industrial – Fishing Industry

- Production Turtle Meat, QSA; Deputy Director of Native Affairs, Thursday Island to Protector of Islanders, Thursday Island, 23 October 1941, SRS505/1 Box 585 Item 4027, QSA.
11 Deputy Director of Native Affairs, Thursday Island to AS, Palm Island, 9 January 1942, SRS505/1 Box 585 Item 4027, QSA; Director of Native Affairs, Brisbane to Manager, Queensland Pastoral Supplies Pty Ltd, Brisbane, 18 April 1944, SRS505/1 Box 823 Item 5623, Correspondence Files (Alphanumeric), Torres Strait – Production – Disposal of Dugong Oil – Sales Southern Firms, QSA.
12 Report Ref. 39/434 Med. & San., 'Dugong oil: its source and therapeutic qualities', c.1951, RSI15058/1 Item 1346, QSA, pp1–3.
13 See the annual reports of C. O'Leary, Director of the QDNA for the years 1951, 1952, 1954 and 1956, published in *QPP*.
14 Deputy Director of Native Affairs to Mrs Lambert, Brisbane, 22 May 1955, SRS505/1 Box 823 Item 5623, QSA; A. D. Love, Manager, Chemist Roush Incorporating Modern Laboratories, 5 February 1958, RSI5058/1 Item 1346, QSA; E. Turner to Director of Native Affairs, Thursday Island, 9 May 1960, RSI5058/1 Item 1346, QSA; Deputy Director of Native Affairs, Thursday Island, Memo, 19 December 1961, RSI5058/1 Item 1346, QSA.
15 Deputy Director of Native Affairs to Chairmen – all Islands, 18 February 1966, RSI5058/1 Item 1346, QSA; Manager, QDAIA, Lockhart Mission to Director, QDAIA, Brisbane, 29 January 1971, SRS505/1 Box 1150 Item 7762, QSA.

Chapter 8

Impacts on whales, sharks and fish

Introduction

The previous two chapters have described historical human impacts on marine turtles and dugongs, which are comparatively well-researched, charismatic megafauna of the Great Barrier Reef and its adjacent waters. In contrast, impacts on some other marine wildlife species in the region are less known. This chapter focuses on historical impacts on some types of whale, shark and fish, although some of those impacts can be reconstructed only partially, due to the paucity of records. While several species of cetaceans are found in the GBRWHA, comparatively little is known about those animals and, even in the year 2000, the population sizes of all species except for humpback whales (*Megaptera novaeanglia*) were unknown (GBRMPA, 2000). This chapter describes the humpback whale fishery in the Great Barrier Reef, for which some documentary evidence illuminates the period 1952–1962. The humpback whales found in the Great Barrier Reef migrate between feeding grounds in Antarctic waters and breeding areas in coastal Queensland; along parts of the Queensland coast, their migration routes bring the animals close to the shore. In those places, the animals are particularly vulnerable to anthropogenic impacts from the adjacent coast. Furthermore, the species is characterised by a long-lived, slow-reproducing life history, with high investment by lactating cows in their calves. Those characteristics impose additional vulnerability on the species in Queensland waters where cows and calves are susceptible to human impacts. GBRMPA (2000) identified numerous anthropogenic impacts on cetaceans, including commercial whaling, harassment, vessel strikes, entanglement in nets, ingestion of litter, underwater explosions, pollution (including noise pollution), disease, live capture and habitat degradation. In addition to historical impacts on humpback whales, this chapter also briefly considers some of the major impacts on sharks and fish in the Great Barrier Reef, although both of those subjects require more detailed, dedicated investigation.

Impacts on humpback whales, 1952–1962

As a result of the large migratory range of humpback whales, the animals found in the waters of the Great Barrier Reef have been affected by historical activities outside the boundaries of that ecosystem. In particular, the commercial humpback whale fishery that operated from 1952–1962, based at the Tangalooma whaling station on Moreton Island, resulted in severe depletion of the species in east Australian waters (Paterson et al., 1994). A brief overview of the impacts of that fishery is provided here. That account belongs in the context of the development of Australian whaling, described for the period between 1791 and 1934 by the marine biologist, W. J. Dakin (1934). That study indicated that an increase in Australian whaling took place between 1800 and 1803; by 1837, the industry had developed into a major industry, based at ports in southern Australia, including Sydney, and carried out by companies such as the South Australian Company and Whale Products Pty Ltd. The earliest operations of the industry were characterised by opportunistic harvesting of whales, and whaling ships sometimes worked the waters of the Great Barrier Reef; however, greater impacts were sustained by the whales of the Great Barrier Reef during their southwards migrations (Jones, 1980; 2002, p87). Scarce documentary evidence illuminates the earliest period of Australian whaling and its impacts; in 1997, Corkeron (1997) reported that the status of cetaceans in the GBRMP remained poorly understood.

The east Australian humpback whale fishery was created in response to an increased demand for whale oil, following the Second World War, and in 1949 the Australian Whaling Commission was formed to co-ordinate Australian whaling, with the intention of generating significant exports of the produce from Australia (Jones, 2002, p87). One company, Whale Products Pty Ltd, was established in 1950 in order to develop the fishery on the east Australian coast. On 1 January 1952, Whale Products Pty Ltd was issued with licences to operate in Queensland coastal waters for a period of five years; the licences permitted that company to kill and process up to 500 humpback whales during the season extending from May until October in each year. The Annual Report for that year by E. J. Coulter (1952, p1012), the Queensland Chief Inspector of Fisheries, stated:

> The first whale was killed on 6 June, and 600 whales were dealt with between that date and 7 October, a permit being given to take an additional 100 whales which had been allotted to another company which did not commence operations.

The animals were processed at the Tangalooma whaling station and at a smaller station in Byron Bay (Figure 8.1). The following year, the quota was increased to 700 animals, which were obtained between May and September (Coulter, 1953, p1016).

In 1954, the quota of 600 animals was again achieved without difficulty; the captures were made between May and September of that year. During that year, Coulter (1954, p1005) reported:

Figure 8.1 A whale captured in east Australian waters, 1950s. Source: Negative No. 43701, Historical Photographs Collection, John Oxley Library, Brisbane

An officer of the Commonwealth Fisheries Office is stationed at Tangalooma during the season to ensure that the provisions of *The Whaling Act* are observed. An officer of the Fisheries Division of CSIRO [the Commonwealth Scientific and Industrial Research Organisation] is stationed there also and collects data and makes observations on the whales that are handled. Such information is summarised in the reports presented each year to the Scientific and Technical Sub-Committee of the International Whaling Commission.

This report is significant as it indicates that the harvest of humpback whales was both legally regulated and scientifically monitored; unlike the previous European fisheries that had taken place in Queensland waters, the operation of the commercial humpback whale fishery was accompanied by the collection of scientific data about the catches.

In 1955, Coulter reported that the east Australian humpback whale population 'appeared to be still in a reasonably stable condition', although a decline in the numbers of that species had been observed by that year in the west Australian fishery. In spite of those reports of over-exploitation of the animals, no reduction in the quota was made; another 600 animals were captured during the season with, as Coulter reported, 'the 1955 catch even showing some improvement'; in 1956, the fifth season of operation of the fishery, 600 humpback whales were caught between June and August (Coulter, 1956, pp1017–18). In 1957, a further 600 animals were caught and processed, between June and August, and Coulter (1957, p1014) stated that CSIRO 'determined that the catch composition of the eastern Australian coast has improved in 1955 and 1956, probably as a result of more careful selection of larger whales'; during that year, the legal regulation of the fishery was also consolidated by means of *The Fisheries Act of 1957*.

However, in 1957, as a result of the scientific monitoring of the harvest undertaken by CSIRO, the lack of ecological sustainability that characterised the fishery was acknowledged, and Coulter (1957, p1014) stated that 'the combined catch of the east Australian humpbacks of the Australian coast and of Antarctic waters will not withstand continuous fishing above the 1956 level'. Nevertheless, no reduction in the quota allocated to Whale Products Pty Ltd was made and another 600 humpback whales were caught between June and August of that year; subsequently, in 1959, CSIRO reported that 'the population of humpback whales along the eastern Australian coast continues to be in a fairly sound condition' and, for the years 1959–1961, the quota was increased to 660 animals (Coulter, 1958, p1015; 1959, pp1096–7; 1960, p1153; Peel, 1961, p750). However, by 1960, changes in the behaviour of the humpback whales were apparent, as Coulter (1960, p1153) stated:

> Recoveries from whale markings show that during the summer of 1958–1959 the eastern population had spread further westwards in the Antarctic than usual, and some mingled and remained with the western population.
>
> The harvests for the years 1959 and 1960 were obtained successfully, although with increasing difficulty, and by 1961 three whaling ships worked to secure the catches. By the latter year, an aircraft was also used to assist in locating the whales.

In 1961, the collapse of the east Australian humpback whale population was acknowledged; A. J. Peel (1961, p750), the Director of the QDHM, reported that only 591 whales of the permitted quota of 660 animals had been achieved by Whale Products Pty Ltd by 30 October, when the season closed, and he stated that the harvest had taken place at an average weekly catch rate of 28 animals, compared with 60 animals during the previous year. A report by CSIRO (cited in Peel, 1961, p751) stated:

> Catch composition studies show that the decline of the population of humpback whales of the western coast continued unchecked during 1960, and that of the eastern coast has begun to decline, although as yet this stock is larger than the remnant of the western coast population.

The following season, only 68 humpback whales were captured between June and August 1962, and the whaling station at Tangalooma ceased operations; by 1963, the infrastructure at Tangalooma station had been sold by Whale Products Pty Ltd and a tourist resort was subsequently constructed in its place (Peel, 1962, p763; 1963, p833).

Hence, the impacts of the whaling industry were severe in Queensland waters, despite the provisions apparently made for the regulation and scientific monitoring of the fishery. The statistics presented above indicate that, during the decade of the operation of the eastern fishery, 6,179 humpback whales were killed at the Tangalooma station. One estimate of the impact of the fishery suggests that around 10,000 humpback whales migrated along the east Australian coast at the commencement of the Tangalooma fishery, in 1952; a decade later, less than 500 animals were thought to survive in that population (Orams and Forestell, 1994). Corkeron (1997) reported that the size of the remaining east Australian population of humpback whales in 1993 was estimated to be approximately 2,500 individuals; he stated that this figure had been achieved after an annual rate of increase in the population of around 10 per cent per year, yet the total estimated population is nonetheless far smaller than the total harvest of the Tangalooma station. Corkeron (1997, p283) also acknowledged that the mortality of the species has also been increased by the activities of illegal Soviet whalers, which he stated have 'killed far more whales than previously thought'. In spite of the regulation and scientific monitoring of the fishery, and its short duration in comparison with the other European fisheries in the region, commercial whaling resulted in a severe reduction in the humpback whale population of the east Australian coast.

Impacts on sharks

Commercial exploitation of sharks in the Great Barrier Reef for the collection of shark fins and the production of shark oil was described in 1890 by Saville-Kent (1890a, p733), who stated:

> At one of the *bêche-de-mer* curing stations in the Great Barrier district, I was informed that a curer had experimentally sent in some dried sharks' fin to Cooktown, and which had readily realised among the Chinese residents a price of no less than 19 d per pound. [...] The livers of sharks [...] yield a valuable oil, while their carcasses, in combination with the waste products from the *bêche-de-mer*, would make excellent manure, akin to guano and particularly rich in phosphates.

Further documentary evidence of the commercial production of shark products was provided by Great Barrier Reef Fisheries Ltd (1929, pp5, 7, 15), which acknowledged the existence of a large market for shark fins, tails, oil, leather, teeth, dried steaks and manure; that company reported that shark fins were sold from between 2s 6d to 10s per imperial pound in China, shark meat obtained £25 per ton and shark leather was used to manufacture shoes, handbags and wallets.

In 1932, the increasing prospects of the Queensland shark fishery were discussed by J. D. W. Dick (1932, p6), the Queensland Chief Inspector of Fisheries, who stated:

> From time to time inquiries are received by the [QDHM] as to sources of shark skins, shark oil, and fins, and there is evidently a growing demand for these products. Some action to test the commercial possibilities of shark products has been taken during the year by a company established at Wynnum, which has also shown a considerable amount of enterprise in the manufacture of edible fish products.

Dick (1935, p1091) later acknowledged that interest in the shark fishery remained high. In 1937, he reported that commercial shark fishing occurred near Bowen, and he stated that those operations were 'in charge of a well-known exponent of that type of fishing, who has had experience in dealing with shark products' (Dick, 1937, p1404). That individual was probably Norman W. Caldwell, a renowned commercial shark fisher employed by Queensland Marine Industries Ltd of Brisbane; the prospectus of that company stated that Caldwell had fished commercially for sharks for many years (Queensland Marine Industries Ltd, 1932, p6).

In the 1930s, sharks were exploited commercially for an increasing range of products. Shark-fin was used to manufacture soups; shark oil was used as a medicine, in the production of cooking oil, in the tanning industry and as a lubricant on ship-ramps; shark meal was used in the manufacture of agricultural fertilisers; and shark leather was sold for the curios trade. Like many other marine resources, the sharks of the Great Barrier Reef were considered to be 'practically inexhaustible'. The operation carried out by Queensland Marine Industries Ltd was based on an average weekly catch of twelve tons of sharks, although a large by-catch of other species also resulted from that operation, and by 1933 another company, Ford Sherrington Ltd, had purchased shark leather from Queensland Marine Industries Ltd for over eighteen months; the leather was produced from the skins of tiger, whaler and nurse sharks (Queensland Marine Industries Ltd, 1932, pp7–8, 11; Goddard, 1933, p221; Roughley, 1936, p252). In one book, entitled *Fangs of the Sea*, Caldwell (1936) referred to the export of hundreds of tons of shark fins annually to China and Malaysia; he also mentioned the manufacture of leather products from shark hides and the production of oil from the shark livers. Another book by Caldwell (1938), *Titans of the Barrier Reef*, contained additional descriptions of the capture of sharks in the Whitsunday Group.

During the 1930s, the fishing of sharks for sport, described by Lamond (1936), had commenced; that activity occurred alongside the commercial shark fishery. Lamond's (1936) account contained evidence of the destruction of tiger and hammerhead sharks using lines. In addition, sharks were destroyed by other fishers who regarded the predators as a nuisance because they interfered with the fish catches (Anonymous, 1933b). One account of 1933, by Northman (1933, p39), described the destruction of sharks while fishing:

> But do not bother too much about killing a shark. Just give him [sic] a bullet, and that will be sufficient. His friends will do the rest. As soon as he is wounded, he is attacked by others.

Several of these popular accounts refer to the occurrence of very large sharks in the Great Barrier Reef; for example, Lamond (1936, p25) described the capture of a sixteen-foot shark, and Caldwell (1936, p91) referred to a hammerhead shark with a hammer six feet across.

Sharks have also been depleted since 1962 as a result of the nets and drum lines set for bather protection. Following two fatal shark attacks in the summer of 1962, A. J. Peel (1962, p764), the Director of the QDHM, stated: 'Tenders have now been called for long-term shark fishing contracts on the South, near North and Cairns coast which are due to commence on 1 November 1962'. From that date until 31 May 1963, Peel (1963, p838) reported, 1,073 sharks and 910 shark pups were captured, figures which 'far exceeded the most optimistic estimates of the probable take'; Peel also stated that 26 grey nurse sharks, which previously had been regarded as rare in Queensland waters, were caught. Catches of similar magnitudes were made for the remainder of that decade. In 1964, of a total catch of 1,056 sharks, Peel reported that most animals were caught in northern coastal waters, with 295 sharks being destroyed in the vicinity of the Cairns beaches alone. By 15 June 1970, a total of 10,622 sharks and 5,643 shark pups had been caught since the introduction of the shark control program in 1962.[1]

Impacts on fish

Fishing has taken place in the Great Barrier Reef since the earliest period of European exploration. A large number of documentary and oral sources describe that activity; those sources exhibit a vast diversity of opinions and perceptions about the nature, methods, extent and impacts of fishing on the resources of the Great Barrier Reef. Furthermore, extensive debates have taken place about the relative impacts of commercial and recreational fishing, and anecdotal reports of decline of fish stocks as a result of both of these layers of fishing effort have been made. Here, those debates are not reviewed in detail, nor are the vast quantities of empirical materials about fishing reviewed systematically; the enormous amount of documentary and oral history evidence that relates to fish species in the Great Barrier Reef precludes exhaustive consideration of the impacts of fishing in this book. Indeed the history of fishing in the GBRWHA requires separate, detailed treatment. Instead, in this section, I present a limited amount of documentary and oral history evidence about selected impacts of fishing, in order to provide an overview of the perceived depletion of fish stocks. The account presented here suggests that the cumulative effect of various impacts on fish may have been significant for some species, and that significant degradation has almost certainly occurred to some fish habitats.

Although fish have been taken from the Great Barrier Reef by Europeans since the arrival of the *Endeavour*, early European fishing took place on an opportunistic basis, and a wide variety of methods were used by fishers. One particularly destructive practice was the dynamiting of coral reefs for fish. Although dynamite fishing is an activity that has been comparatively overlooked in accounts of the history of the Great Barrier Reef, documentary and oral history evidence indicates that the practice was once prevalent in Queensland coastal waters and reefs. In 1913, the Queensland Treasury Departmental Committee investigated the Queensland fisheries; that Committee commented that, at almost every port, 'complaints were made that dynamite is freely used for taking fish' (Boult et al., 1913, p1041). The Committee stated that:

> The use of explosives for the purpose of obtaining fish in the inland waters has, it is stated, been most freely adopted in the waters in the neighbourhood of any large construction works which have been carried out, and to this abuse the residents attribute the scarcity of fish owing to the destruction of so much of the 'small fry'.
>
> (Boult et al., 1913, p1052)

In an attempt to control the problem of dynamiting, prosecutions for the illegal use of explosives for taking fish were made in 1925, in the Brisbane and Maryborough districts, and large fines were issued. Describing those measures, the Director of the Queensland Marine Department stated: 'It is hoped these will have a salutary effect on persons disposed to this method of destroying fish, which is most wasteful to fish life and dangerous to the user' (Forrester, 1925, p295). Numerous prosecutions for the use of explosives for fishing were reported in Queensland during the period 1925–1970.[2]

However, preventing the use of dynamite by fishers was not easy. In 1931, J. D. W. Dick (1931, p6), the Acting Chief Inspector of Fisheries, reporting a prosecution for the use of explosives, stated that:

> This nefarious practice is particularly destructive of young fish, and is most difficult to detect, as the offender can carry the necessary equipment in his [sic] pocket, and usually selects some infrequented locality in which to carry out his purpose.

By 1933, despite regulations and publicity aimed at preventing the use of explosives, the practice had not ceased; J. Wyer, the Honorary Secretary of the North Queensland Naturalists' Club (NQNC), stated that 'dynamiting on the reef is as prevalent as ever', a fact he attributed to inertia on the part of those who were responsible for enforcing the legislation. As a result of dynamiting for fish, Wyer stated: 'The amount of damage in the aggregate is enormous and every effort should be made to bring the offenders to book'.[3] In 1937, the Honorary Inspector of Fisheries at Green Island acknowledged similar problems at Green Island, where it proved difficult to control 'this popular fishing ground'. That observer stated: 'Dynamiting of fish in the past has been prevalent along the reef, and from Fitzroy Island and Oyster Cay a distance of 20 miles should be visited at least once weekly by an Inspector of Fisheries'.[4] In addition to those documentary sources, two oral history informants recalled instances of people fishing using dynamite in the Cairns area, although one indicated that the practice became less common after the Second World War.[5] By that date, however, dynamiting for fish had taken place in Queensland coastal waters and reefs for more than three decades.

In general, in the early period of European settlement in Queensland, fishing was restricted to the coastal zone, since plentiful catches could be obtained from the shore without the use of boats. However, the twentieth century has been characterised by an overall dramatic increase in fishing effort in the Great Barrier Reef, and more systematic manipulation of fish stocks had commenced by the 1920s; in 1925, for example, Taylor (1925, p217) referred to fish breeding in the Great Barrier Reef. Yet during the first half of the twentieth century, the pelagic fisheries of the Great Barrier Reef were assumed to be secure and many accounts describe the abundance of fish that was available. For example, one account (Anonymous, 1929) stated that, at Lindeman Island:

> Mackerel or king fish trailing during the cooler months is unsurpassed in any part of Queensland. Fifteen mackerel have been landed in one hour, each weighing from seven to thirty-five pounds, with the spinner. [...] An abundant supply of fresh fish for the table is to be obtained with net and trap.
>
> During the 1920s, the tourist resorts of the Great Barrier Reef were promoted particularly on the basis of the high quality of recreational fishing at popular islands.

The availability of good fishing was described by several documentary sources in 1933 (Northman, 1933; Reid, 1933). One account considered that the waters of Queensland contained at least one thousand fish species, and that more were being discovered yearly: the best-known fish included the snapper, swordfish (*Tetrapterus*), groper (*Epinephelus lanceolatus*), bream, whiting, mullet, mackerel, flathead, tailor, jew-fish, trevally, emperor, clupeid, sole and flounder (Tilghman, 1933, pp63–4). Despite that abundance of species, Stoddart (1933, pp217, 219–20) claimed that Australia imported about one third of a million cwt of fish, annually, at a cost of around £1,700,000; he advocated the expansion of the Great Barrier Reef fisheries instead. In particular, Stoddart (1933) acknowledged the potential of the area between Townsville and Bowen, which was characterised by a muddy bottom, and he argued that this area could support rich trevally and prawn fisheries; in particular, large tiger prawns were found in this area. In contrast, Stoddart (1933) suggested that the area between Townsville and Cairns was unsuitable for trawling, but that area could instead be exploited for the profitable king snapper (*Lutianus sebae*), kingfish and mackerel. Stoddart (1933) reported that, further north in the Great Barrier Reef, enormous shoals of Murray Island sardines could be found throughout the year, and that mullet and garfish were available throughout the reefs.

If by 1933 the commercial fisheries of the Great Barrier Reef remained under-exploited, the possibilities offered by game fishing were already being explored, and Reid (1933, p39) reported that increasing numbers of tourists from New South Wales and Victoria were visiting the Great Barrier Reef to pursue that activity. An advantage of the Great Barrier Reef for this sport was that it contained sheltered fishing grounds, close to continental islands, in which game fishing could be carried out in poor weather. Tilghman (1933, p223) reported that game fishing took place from Lady Elliot Island to Torres Strait, and the sport was concentrated in the vicinity of coral reefs; the base of the activity, however, was located at Hayman Island, where very large catches were obtained. Tilghman (1933, p225) stated:

> During the winter season these two fishing skippers [Bert Hallam and Boyd Lee] have each taken enough big mackerel in a morning to fill the ton ice boxes in their launches. They used three heavy hand lines of course, and anglers would view such fishing as slaughter, but it shows the fish there are.

The game fishermen landed very large specimens; swordfish, blue pointer sharks, leaping-tuna and giant turrum were sought in the waters near the Whitsunday Islands and the islands offshore from Gladstone and Bundaberg. Large stingrays, such as the animal shown in Figure 8.2, were also sought by game fisherman, such as an 800-pound specimen that Tilghman (1933) reported was caught near Hayman Island.

During the 1930s, with increasing access to the resorts of the Great Barrier Reef – and with increasing boat ownership, which allowed fishing parties to have direct access to the reefs – the incidence of fishing increased. Some popular books describing fishing in the Great Barrier Reef were published, and several individuals became celebrities as a result of their promotion of fishing near the reefs and cays, including Zane Grey, Bert Hallam and Boyd Lee; Caldwell (1938, p23) referred to Boyd Lee as a 'Barrier Reef celebrity' and a professional fisherman. Caldwell (1938, p90) also described a seasonal variation in the most abundant fishing. He stated:

> Winter months are [north Queensland's] great harvest time, when the immense schools of striped tuna, locally called 'kingies' (kingfish) work into the warmer waters from the south. The boats bring back huge hauls of this excellent table-fish. Townsville and Cairns absorb large quantities, the rest being railed to the Brisbane markets.

Figure 8.2 A large stingray captured in the Great Barrier Reef, c.1930. Source: Negative No. 44419, Historical Photographs Collection, John Oxley Library, Brisbane

In part, however, the success of the fishermen could be attributed to fishing methods which are now regarded as unsustainable, such as the blocking of creek mouths using fishing nets, which was a practice recommended by Boyd Lee. Another destructive practice was the shooting of fish in rocky pools on coral reefs, using pea-rifles, which was advocated by Northman (1933, p39) at Magnetic Island.

By 1939, the resources of the Great Barrier Reef fisheries were no longer perceived as being unlimited, and restrictions on the pelagic fisheries had been introduced. On 20 July 1939, an *Order in Council* prohibited 'the taking of all or any kind of fish as defined by those Acts in any Queensland waters specified in the Order'; those restrictions were enacted under *The Fish and Oyster Acts, 1914 to 1935*.[6] Subsequently, reports of damage to fish populations were received by the Queensland Department of Fisheries. One example of destruction concerned the sardine stocks at Green Island, which were reported to be experiencing increasing pressure as the number of visitors to the resort increased. In 1941, the lessees of that cay, Hayles Magnetic Pty Ltd, complained about the destruction of sardines at Green Island.[7] The company stated:

> We would like to stop the destruction of sardines which visit the shores of Green Island in large shoals, and [are] an attraction for tourists. It has been known people throwing cast-nets from the jetty will catch ¾ of a sugar bag of sardines in one cast. After this happening for a month or so the sardines disappear for from three to twelve months. We would not like to stop people securing sufficient bait for fishing while they were on the Island, but to fill cases and bags of sardines to take away should be stopped.[8]

The Queensland Chief Inspector of Fisheries responded to this complaint, reiterating that the use of cast nets was illegal throughout Queensland; the caretaker at Green Island, W. D. Scott, who was also a Queensland Honorary Inspector of Fisheries, was instructed to seize cast nets under Section 44 (1) (d) of *The Fish and Oyster Acts*.[9]

If the increasing exploitation of the Great Barrier Reef fisheries had become a cause for concern by the end of the 1930s, the intensity of fishing in these waters nevertheless increased in the subsequent decades: that increase was a consequence of the economic importance of the commercial fisheries to the development of the State. In addition, the significance of charter operations for fishing parties increased. For example, the charter fishing industry expanded during the 1960s, stimulated by the growth of the black marlin (*Makaira indicus*) fishing industry; that industry commenced in 1966 in the waters between Cairns and Lizard Island (Dodds, 2004).

Another type of fishing that grew in popularity between 1930 and the 1960s – with severe consequences for reef fish populations – was spearfishing. This activity was controversial since it was regarded as an easy method of wreaking significant destruction on large fish; nevertheless, spearfishing was pursued at many resort islands and reefs in the Great Barrier Reef. As recently as 1969, complaints were received about the damage caused by spearfishers to coral reef fish, although the problem was difficult to address since the offshore reefs lay outside the jurisdiction of the Queensland authorities; in any case, supervision of the more remote reefs was difficult.[10] Some evidence of the impacts of spearfishing at Heron Island was provided in the following account of 1969 by Peel, the Director of the QDHM, who stated that:

> Prior to June 1963, spear-fishing was permissible in the vicinity of Heron Island. It was then claimed by the management of the tourist resort, and supported by the Great Barrier Reef Committee, that it was no longer possible for tourists in the glass-bottomed boats to view larger groper: as fast as such fish were located, they fell prey to spear-gunners.[11]

Damage also occurred at Green Island where, in 1972, objections were raised to the 'wilful destruction of coral and the use by spear fishermen of bullets fired underwater', as the Hon. J. Herbert, Queensland Minister for Labour and Tourism, acknowledged.[12]

In a letter describing the destruction caused by spearfishers at Lizard Island, in 1974, the coral reef photographer, Roger Steene, wrote:

> I have been a constant visitor to the Island and its adjacent reefs for the past 18 years and environmental changes seem to be ever increasing since the establishment of the aerodrome and the discovery of the island as a good anchorage and camping area. [...] On the reef at North East Point, a colony of large coral trout, *Plectropoma maculatum* lived for many years. These were huge specimens, 30–40 lbs and it was the only part of the Great Barrier Reef where I had seen ten or more big specimens living together in such a tiny area.
>
> These fishes were the subject of many photos as they were unique due to the fact that I had tamed them as they would readily approach a diver to collect food prepared for them. This colony of fishes has entirely disappeared in the last couple of years. A camp was made nearby for several months by a group making a film and conversation led me to believe they had initiated the killing of these unusually tame fishes. I believe a great opportunity has now been lost to study such magnificent specimens in their natural habitat. I partially blame myself for feeding them and winning their trust so as to make them un-missable targets for spear-guns of irresponsible people.

The impacts of spearfishers were pronounced since the largest fish were targeted by the divers; this concentrated the mortality among the larger species of reef fish, and among the larger individuals within populations, leading to a progressive reduction in the size of fish apparent on popular coral reefs.[13]

Another fishing activity that has caused degradation in the Great Barrier Reef is the collecting of aquarium fish. Steene reported that at Lizard Island, during the same visit, an abundance of valuable species of aquarium fish was found. He stated:

> In my present study of *Chaetodontidae* and *Pamacanthidae*, I made an exciting and important discovery during my recent stay. In shallow, protected water in the lagoon, I located an area which apparently is a prolific breeding ground for several rare species of fishes. This included different types, but most importantly, I found fourteen juveniles of *Heniochus singularius* a species which was previously known from only two specimens in Australian waters.
>
> Similarly *H. monocerus* and *C. bennetti* were abundant in juvenile form. Neither of these is a common species and would be considered a 'gem' to aquarists. In the same locality, I also found and photographed a species of *Solenichthys* that is so rare that only four specimens are held in museums in the world. This also would be a 'gem' for an aquarium and would fetch a handsome price.
>
> With the ever increasing popularity of salt water aquariums in the home, I consider the lagoon at Lizard a 'golden egg' for professional aquarium fish

collectors. Since my return to Cairns, I have met two different fish collecting parties, apparently with big budgets and well organized, who plan working northern waters making Lizard Island their base.[14]

However, several oral history informants reported that as a result of the collection of coral reef fish for the aquarium trade, the abundance of the most popular fish has declined.[15]

The decline of many of the commercial and recreational fisheries of the Great Barrier Reef as a result of over-exploitation of fish has been described by Love (2000, p98), who stated:

> Live coral reef fish, from the Great Barrier Reef and Pacific Island nations, are air-freighted to Asia for restaurants, where they fetch large prices as status symbols. In 1995 the live fish trade brought 25,000 tonnes of live coral reef fishes into Southeast Asia. In 1996, a live groper sold in Hong Kong for US$10,256.

Numerous oral history informants provided additional anecdotal reports of overfishing and the depletion of fish stocks in the Great Barrier Reef; those reports, in addition to the documentary evidence presented above, suggest that impacts on fish in the Great Barrier Reef since European settlement may have been considerable and should ideally be the subject of separate, dedicated investigation.[16]

Summary

This chapter has presented evidence of historical impacts on some types of whale, shark and fish, although that evidence is partial and there is scope to investigate impacts on sharks and fish, in particular, in much greater detail. Nevertheless, the account presented here suggests that some significant transformations have occurred in some of the whale, shark and fish populations of the Great Barrier Reef. Severe depletion of east Australian humpback whales occurred between 1952 and 1962 due to the operation of the commercial whaling station at Tangalooma, leading to the collapse of that fishery due to over-exploitation of the resources on which it was based. That episode contains a salutary lesson for the management of marine living resources, for the collapse of the humpback whale fishery occurred abruptly and with little warning, despite the existence of government regulation and scientific monitoring intended to ensure a sustainable harvest. Although that commercial fishery has now long ceased, humpback whales still face many other threats, including vessel strikes, harassment, entanglement in nets, ingestion of litter and pollution (including noise pollution).

Yet while the historical impacts on east Australian humpback whales are reasonably well-documented, some of the other impacts mentioned in this chapter – those affecting sharks and fish in the Great Barrier Reef – are less well-described. Nevertheless, there is evidence to suggest that the actions of shark fishers (including the capture of sharks for their fins), dynamite fishers, spear-fishers and other commercial and recreational fishers have significantly depleted some types of sharks and fish in the Great Barrier Reef, and in some cases have also directly degraded reef habitats, with unknown ecological consequences. Those impacts were noticed by many visitors to the Great Barrier Reef, in terms of reductions in both the availability and the size of fish caught, and the over-exploitation of sharks and fish in the region extends far back in the historical record, at least to the period of the earliest development of tourism in the region. Human impacts on both sharks and fish in the Great Barrier Reef, therefore, deserve more detailed, specialist investigation.

Notes

1. See the statistics published in the Annual Reports, QDHM, *QPP*, 1963–1970.
2. See the Annual Reports, QDHM, *QPP*, 1925–1970.
3. In-letter, J. Wyer, Honorary Secretary, NQNC to Honorary Secretary, GBRC, 20 September 1933, PRV8340/1 Item 1, QSA.
4. In-letter, M. T. Keating, Honorary Inspector of Fisheries, Green Island to Chief Inspector of Fisheries, Brisbane, 21 November 1937, PRV8340/1 Item 1, QSA.
5. OHC 17, 2 September 2003.
6. *Order in Council*, Brisbane, 20 July 1939, PRV8340/1 Item 1, QSA.
7. In-letter, Secretary, Queensland Sub-Department of Forestry, Brisbane to Secretary, QDHM, Brisbane, 19 December 1941, PRV8340/1 Item 1, QSA.
8. In-letter, Secretary, Queensland Sub-Department of Forestry, Brisbane to Secretary, QDHM, Brisbane, 5 September 1941, PRV8340/1 Item 1, QSA.
9. Out-letter Ref. 3740, Queensland Chief Inspector of Fisheries, Brisbane to Inspector of Fisheries, Cairns, 11 September 1941, PRV8340/1 Item 1, QSA.
10. Out-letter Ref. 9-1-17, A. J. Peel, Director, QDHM to Under-Treasurer, Brisbane, 17 March 1969, SRS31/1 Box 33 Item 573, General correspondence files – Licences to remove coral, QSA, p1.
11. Out-letter Ref. 9-1-17, A. J. Peel to Under-Treasurer, Brisbane, 14 February 1969, SRS31/1 Box 33 Item 573, QSA, p1.
12. Hon. J. Herbert, Queensland Minister for Labour and Tourism, Brisbane to Hon. V. B. Sullivan, Queensland Minister for Lands, Brisbane, 22 February 1972, SRS5416/1 Box 66 Item 447, NP836, Trinity 'R' – Green Island – Protection of Marine Life, QSA.
13. R. Steene, Cairns to S. Domm, Director, Museum Research Station, Lizard Island, 20 May 1974, p1, SRS5416/1 Box 28 Item 179, NP153, Flattery 'A' – Lizard Island, QSA.
14. R. Steene, Cairns to S. Domm, Director, Museum Research Station, Lizard Island, 20 May 1974, pp1–2, SRS5416/1 Box 28 Item 179, NP153, Flattery 'A' – Lizard Island, QSA.
15. OHC 21, 10 September 2003; OHC 31, 4 October 2003.
16. OHC 5, 11 February 2003; OHC 6, 17 February 2003; OHC 8, 27 February 2003; OHC 21, 10 September 2003; OHC 23, 15 September 2003; OHC 29, 24 September 2003; OHC 40, 12 November 2003.

Chapter 9

The impacts of coral and shell collecting

Introduction

This chapter examines the impacts of coral and shell collecting in the Great Barrier Reef, activities that have been comparatively neglected in accounts of the history of the ecosystem, yet which were widespread, cumulative and probably severe for some species. Coral and shell collecting occurred in more places, and for longer periods, than has previously been documented. Four main types of coral and shell collecting occurred in the Great Barrier Reef: informal collecting, scientific collecting, unregulated commercial collecting and licensed collecting. Although individual occurrences of coral and shell collecting were comparatively small and localised in the context of the scale and diversity of coral reefs, the cumulative impact of many collectors, in many places, over a long period of time is likely to have been considerable. In particular, at major tourist centres – such as Hayman, Heron and Green Islands – the degradation of coral reefs was probably severe, with the result that some parts of the Great Barrier Reef were 'loved to death' by visitors.[1] In addition to the informal removal of coral and shells by visitors to the Great Barrier Reef, commercial coral and shell collecting has been a consistent impact on numerous reefs.

The reconstruction of informal and commercial coral and shell collecting in the Great Barrier Reef is difficult for several reasons: the lack of systematic records; the difficulty of estimating harvests due to illegal collecting; the problems in identifying species; the limited extent of regulation and monitoring of the activities of collectors; the vast geographical range of the reefs on which collectors worked; and the reluctance of some coral and shell collectors to contribute oral history evidence of their activities. In addition to those problems, individual instances of coral and shell souveniring have been regarded as trivial, and the changes that resulted from souveniring were generally imperceptible because they occurred gradually. Nonetheless, the account presented below outlines the general scope of coral and shell collecting, and it provides several examples of collecting activities and their impacts in particular locations.

Coral harvesting in Queensland has been regulated since 1933, when the first restrictions on the removal of coral from foreshores in Queensland were introduced, by which time an industry had formed to supply the souvenir market (Harriott, 2001, p11).[2] Analysis of documentary records reveals that informal coral collecting pre-dated the regulation of that fishery in Queensland; that activity was both intensive and sustained at many locations in the Great Barrier Reef. Evidence of the scale of the commercial coral collecting industry is found in the records of the coral licences that were issued to professional collectors; the surviving licences are held at the QSA. Some oral history sources, historical books and photographs supplement those records with additional details of the extent of coral collecting and its impacts. However, in the surviving records of coral and shell-grit licences issued by the QDHM, uncertainty exists about the precise use for which the permits were intended. The sequence of licences is continuous with the licences that were issued for coral mining (see Chapter 11), which initially took place for the manufacture of agricultural and industrial lime; however, terrestrial sources of agricultural and industrial lime eventually replaced lime manufactured from coral, and coral collected since the 1950s increasingly supplied the curios and ornamental trades instead.[3]

Coral and shell collectors probably degraded reefs in similar ways, since they removed target species from reefs and also damaged corals in the process by reef-walking (see Chapter 12). Coral and shell collecting form part of a group of harvesting activities that have depleted many reef organisms, and to some extent the distinctions between those activities are artificial; many collectors collected a range of marine specimens of different types during their visits to the Great Barrier Reef. Overall, the evidence presented in this chapter suggests that significant, prolonged and widespread removal of both coral and shells has occurred in the Great Barrier Reef since the earliest European exploration of the ecosystem took place.

Coral collecting in the Great Barrier Reef

Early instances of the ornamental use of coral date at least to 1879, and the collection of coral from the Great Barrier Reef for curios has taken place continuously throughout the period of European settlement. In addition to the collection of coral for curios and souvenirs, coral was also collected for scientific investigations by early European explorers and naturalists since those aboard the *Endeavour* in 1770. For instance, describing the scientific apparatus aboard the *Endeavour*, one correspondent wrote to the renowned taxonomist, Carl von Linné (Linnaeus):

> No people ever went to sea better fitted out for the purpose of Natural History. They have got a fine library of Natural History; they have [...] all kinds of nets, trawls, drags and hooks for coral fishing; they have even a curious contrivance of a telescope, by which, put into the water, you can see the bottom at great depth.
>
> (cited in Beaglehole, 1955, p cxxxvi)

Coral collecting was a feature of many subsequent European exploratory voyages, including those of Joseph Beete Jukes (1871, p230), who discussed his own coral collection in a letter of 27 July 1844:

> How you would envy the corals which we get here! The most magnificent masses of branched corals are now dying on the poop; but, alas, they are too bulky and too brittle to get home, so I shall content myself with small pieces.

Other instances of coral collecting by early European explorers, naturalists, natural historians and scientists were described by Bowen and Bowen (2002), who showed that large collections of coral were transported from the Great Barrier Reef to institutions in Sydney and London. Oral history evidence also suggests that large scientific coral collections were created before 1960, including a large collection made during a voyage aboard the *Cape Moreton* by Professor Stephenson of the University of Queensland and Dr Wills of Cornell University, and another collection made during the scientific expedition to Low Isles in 1954, although those collections were not maintained.[4]

However, those collections were few in number and highly selective; they formed a very small part of the cumulative impact of coral collecting. In contrast, the collection of coral for commercial ventures represented a much more significant impact on coral reefs. The coral trade had commenced by 1879, when six packages of coral were exported from Queensland to New South Wales.[5] In 1890, Saville-Kent (1890a, p734) stated that:

> A remarkable species of coral that is not infrequently obtained by the pearl-shell divers in Torres Straits and throughout the Barrier region is the black coral, *Antipathes arborea*. This coral possesses a high commercial value in the Indian market, the supplies hitherto having been chiefly derived from the vicinity of Jeddah, in the Red Sea. I am informed that the produce of the Jeddah Fishery has greatly diminished within the last few years, and that the discovery of new sources of supply would be gladly welcomed. There is, I consider, every element in favour of the development of a profitable black-coral fishery in North Queensland waters.

By around 1900, coral collection was taking place at Masthead Island, as Figure 9.1 illustrates, and by 1929 the commercial collection of coral – including other species besides *Antipathes arborea* – for sale as curios and ornaments had increased. An account of Green Island produced by the Cairns Harbour Board stated: 'There is a caretaker on the island who has a very fine exhibition of reef products and marine life, and pretty coral specimens are obtainable at a very low cost' (Cairns Harbour Board, 1929, p46). In addition, visitors to the island were encouraged to explore the reef at low tide for themselves, and the opportunity to collect coral souvenirs was regarded as one of the attractions of the island resorts.

Figure 9.1 Coral collecting at Masthead Island, c.1900. Source: Negative No. AP3:433, Robert Etheridge Photograph Collection, Australian Museum Archives, Australian Museum, Sydney

However, the activities of tourists – in particular, taking coral from the reefs – caused concern about environmental degradation at the major resorts, including Hayman, Heron and Green Islands where, from 1930 onwards, coral specimens were readily available as ornaments and curios (Barrett, 1930, p375). One account of coral collecting at Hayman Island in 1932 and 1933, for instance, written by Marks (1933, pp14–15), stated:

> The excitement was intense as, again and again, the diver who accompanied us returned to the surface grasping bunches of living jewels plucked from the depths. Soon the boat was full of pieces of coral of different shapes and sizes, but the illusion was lost. The dainty fairy-like forms proved to be mere slimy skeletons, and the colour blurred and faded as the living organisms within, away from their natural element, gradually succumbed to the heat of the tropical sunshine, and began to give out a most offensive odour. But even in death the fragments retained their beauty of form, and, cleaned and bleached, were eagerly sought after as souvenirs. To return from the Great Barrier Reef without some specimens of its coral framework was unthinkable!

The corals collected during that particular expedition included 'mushroom' and 'brain' corals, which were removed from the reef using pick-axes.

Concerns about the damage to coral reefs caused by this type of souveniring – and the likelihood of anticipated additional damage as the tourist industry developed – were expressed, as the following account by the Secretary of the Provisional Administration Board of the QDHM illustrates:

> A suggestion was made to this Department by the Director of the Queensland Government Tourist Bureau that it is desirable to prohibit or restrict the removal of live coral from Queensland waters, in view of anticipated developments of the tourist traffic to islands in the Barrier Reef area and the possibility of considerable destruction of growths of coral forming scenic attractions in the neighbourhood of the tourist resorts.[6]

Nevertheless, some degradation of coral reefs had already occurred, the Secretary reported: for example, in 'the Stone Island area where tourists and others have done some damage to the coral formations from a scenic point of view'.

Another area about which early concerns about the coral collecting were publicised was the Whitsunday Islands; one account, written by H. G. Lamond in 1933, requested the Queensland government to prohibit coral collectors 'from removing oysters or coral, shells and other beauties from the Molle reefs' since degradation was occurring in those places. In another letter, Lamond argued that damage to the reefs was occurring, not only as a result of the removal of specimens, but also because other corals were damaged in the process. In the same year, the Queensland Government passed legislation to protect the most vulnerable locations by prohibiting the taking of coral from the foreshores and reefs of eighteen islands: Masthead, Heron, Lady Musgrave and North West Islands (Bunker Group); Middle and South Islands (Percy Isles); Tern and Red Bill Islands (Northumberland Islands); Scawfell, Molle, Shaw, Lindeman, Hayman, St Bees and Brampton Islands (Cumberland Islands); Stone Island (Edgecombe Bay); Bait Reef; and the foreshores and reefs of Cid Harbour (Whitsunday Island).[7] The locations of those early protected areas are shown in Figure 9.2. In 1937, the foreshore and reef surrounding Hamilton Island (Whitsunday Passage) were added to the list of protected areas, followed by other foreshores and reefs in the Whitsunday and Cumberland Islands, in 1939.[8]

Nevertheless, coral collecting remained a popular activity amongst both amateur collectors and naturalists. Ellis (1936, p83) described the attraction of coral collecting as follows:

> So far as naturalists are concerned, I can hardly imagine one being happier than when taking a stroll at low spring tide on the Barrier Reef, with its wealth of shells, corals, crabs, sea-urchins, *bêche-de-mer*, and other strange things that only a naturalist could classify. Every stone one turns over reveals material for a collection; every piece of live coral broken off seems to add its share; not only the polyp which made the structure, but the weird and wonderful tiny crabs, shrimps, and little fish that make their homes among the branching coral. Everything seems to be teeming with life. And it is not necessary to be a naturalist to enjoy these wonders; anyone with a love of nature would be thrilled. The scale on which things are done, too, is befitting the noble proportions of this great reef.

The impacts of coral and shell collecting 133

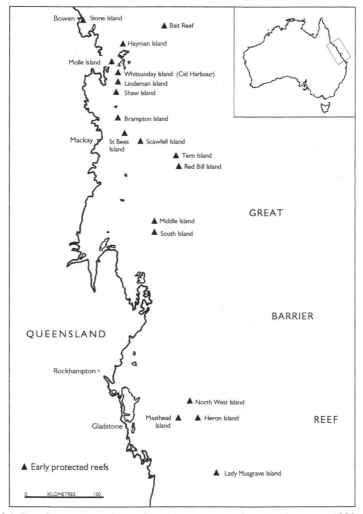

Figure 9.2 The first coral reefs and foreshores protected by legislation, in 1933. Source: Author, compiled from information provided in Order in Council, 1 June 1933

In contrast, Ratcliffe (1939, p139) described the disappointment he experienced when walking across a dead coral reef between Dunk Island and a smaller island of the Family Group, at low tide, and finding few biological specimens. Yet the practice of coral collecting and the treatment of specimens had by then become well-established; an efficient method of cleaning coral by covering it in coral sand for about a week was commonly practised, and coral specimens were then tinted in an attempt to reproduce the colours of the living reef (Richards, 1937, p73).

In spite of the legal protection of some coral reefs that existed since 1933, complaints were made about the degradation of reefs by coral collectors. As Bowen and Bowen (2002) acknowledged, early conservation concerns had already been expressed by representatives of the Great Barrier Reef Committee (GBRC). The GBRC was founded in 1922, largely as a result of the efforts of Henry Richards, Professor of Geology at the University of Queensland, to promote the systematic scientific investigation of the Great Barrier Reef. The GBRC had two explicit aims: to investigate the nature and formation of the Great Barrier Reef, and to develop fuller knowledge of the development and growth of the products of great economic value in the Great Barrier Reef 'so that the Commonwealth may use them in the most efficient and wealth-producing manner' (Bowen and Bowen, 2002, p237). The formation of the GBRC followed the Pan-Pacific Scientific Conference of August 1920, at which the need for a marine biological survey of the Great Barrier Reef was acknowledged. In 1920, Richards had written to Sir Matthew Nathan, Governor of Queensland, outlining a programme for Great Barrier Reef research, including studies of its economic potential. He also argued for an investigation of the economic resources of the Great Barrier Reef, stating that 'the exploitation of the economic wealth of the Great Barrier Reef has gone on and we stand idly by' (cited in Bowen and Bowen, 2002, p234–5).

Richards' views were shared by the eminent reef scientist, Charles Hedley, of the Australian Museum in Sydney, who argued for the conservation of natural resources and who warned about the dangers of uncontrolled exploitation of the Great Barrier Reef. In part, however, these early expressions of conservation thought were motivated by patriotism and a desire to regulate the activities of foreign harvesters. Concerns had already been expressed about the degradation of the Torres Strait pearl fisheries; increasingly, other concerns were expressed that Government fisheries regulations allowed the exploitation of marine resources by 'vagrant licensees' until the resource base collapsed. Hedley advocated the introduction of 'a patriotic policy' with the aim of replacing the 'wandering and foreign population which subsists on our marine tropical products, by resident European fishermen' supported in turn by the advancement of science through zoological research and legislative protection (cited in Bowen and Bowen, 2002, p234). Hence, during that period, early conservation thought developed due to motives that were largely protectionist and oriented to the economic development of Queensland. The GBRC acknowledged that some of the economic resources of the Great Barrier Reef – in particular, pearl-shell and *bêche-de-mer* – had already been heavily exploited; other resources, including 'certain species of reef corals for bleaching, painting and sale as curios' and 'other kinds of decorative soft corals', required increased protection from casual depredation (cited in Bowen and Bowen, 2002, p239).

The increasing engagement of the GBRC with conservation during that period can be understood in the context of the wider development of the conservation movement in Queensland. Following a period of unchecked expansion and exploitation of natural resources during the nineteenth century – and as a result of which the complete destruction of some shell and timber resources occurred — the Queensland Government had by 1900 commenced designating small land areas as national parks. The dominant drive to clear the land for agriculture had caused the widespread destruction of forests and wildlife; the pace and extent of that land clearance led, in turn, to controversies about the need for protection of terrestrial resources (Powell, 1988). In contrast to terrestrial environments, however, the Great Barrier Reef remained virtually inaccessible to the public and a low priority for the Government. While the GBRC advocated the scientific investigation and economic exploitation of the Great Barrier Reef, it increasingly faced political pressures and protests about the impacts of those activities. In particular, many complaints were made about the plunder of reef resources — especially dugongs, trochus and *bêche-de-mer* — by Japanese crews. Some charismatic figures, such as the 'beachcomber', Edmund Banfield, and the marine cinematographer, Noel Monkman, expressed concerns about the degradation of the Great Barrier Reef resulting from activities such as mining, animal harvesting and resort development.

By 1930, the use of the Great Barrier Reef had increased considerably as its waters had become a major shipping route. Reef tourism also expanded during the 1930s; that industry was promoted after 1930 when the Mackay Chamber of Commerce initiated the development of Lindeman Island as a tourist resort. The Queensland Government Tourist Bureau (QGTB) was created as a publicity and booking office, and was functioning in both of those roles by 1932. The main strategy of the QGTB was to attract visitors from Southern states to Queensland, and the organisation launched a promotional literature campaign to highlight the attractions of the Great Barrier Reef, including its 'economic possibilities' (Bowen and Bowen, 2002, p290). By 1935, a major campaign to attract tourists to Queensland was in progress, as reflected in works such as *On the Barrier Reef* (Napier, 1928) and *Wonders of the Great Barrier Reef* (Roughley, 1936). While the Great Barrier Reef received increasing numbers of visitors after 1930, however, the lack of a single co-ordinating authority to manage the Great Barrier Reef led to isolated responses to instances of degradation, and to several authorities administering various parts of the ecosystem in response to varying pressures and concerns. Yet an overall strategy for protecting the marine resources of the ecosystem was lacking, and the GBRC attempted to persuade the Queensland Government of the need for at least a basic level of regulatory control over the Great Barrier Reef.

Significant conservation initiatives and legislation were introduced in Queensland during the 1930s. In 1930, the National Parks Association of Queensland (NPA) was formed, followed in 1932 by the NQNC, in Cairns. In 1933, the protection of wildlife on some islands of the Great Barrier Reef was increased when several sanctuaries were proclaimed under the *Animals and Birds Act, 1921*. Increased protection for various marine parts of the Great Barrier Reef was afforded by means of a series of *Orders in Council*: in 1933, 1937 and 1939, the removal of coral from many reefs and foreshores was prohibited; in

1935, all the islands of the Great Barrier Reef were declared wildlife sanctuaries; and, by 1939, one hundred islands had been designated as national parks. Those regulations were enforced by Honorary Inspectors given legal powers under the *Fish and Oyster Acts*, and Honorary Rangers under the *Fauna and Native Plant Protection Act*. The QGTB also advised visitors about regulations and encouraged compliance with their provisions. The introduction of those initiatives and regulations suggests that, during the early 1930s, both increased visitation and use of the Great Barrier Reef and wider acknowledgement of its economic potential prompted some efforts to afford its resources a minimum standard of protection.

In 1938, E. O. Marks, the Honorary Secretary of the GBRC, wrote to the Queensland Treasurer, stating:

> This Committee has for many years felt much anxiety in regard to the harm which must result from promiscuous gathering of marine and other trophies, and thoughtless destruction of fauna and flora along the Queensland coast. The effects of such vandalism are necessarily greatest in the most accessible places – especially in the vicinity of tourist resorts.[9]

The degradation was of particular concern in the Whitsunday region; another report, by the lessee of South Molle Island, Mr A. W. Bauer, claimed that 'the coral reefs surrounding Mid Molle and Denman Islands are suffering through the action of persons removing coral'. The Director of the QGTB suggested that those two islands should be given the same legal protection from coral collectors as other protected reefs. In July 1939, the number of foreshores and reefs protected under *The Fish and Oyster Acts, 1914 to 1935* was increased to include the remainder of Whitsunday Island as well as Mid Molle, Denman, Hook, Border, Deloraine, West Molle and Long Islands in the Whitsunday Group, and Seaforth Island in the Cumberland Group.[10]

Yet the legal protection of coral reefs did not prevent their degradation by coral collectors, who continued to souvenir specimens illegally. The attractions of 'reefing' were described by the Secretary of the Queensland Office of the Commissioner for Railways who, after visiting Lady Musgrave Island, wrote that:

> Lady Musgrave has extensive coral reefs which provide ample opportunities for reefing at low tides when tourists can see every variety of marine growth and life. [...] On the edge of the reefs and in coral pools, coral gardens flourish in all their beauty.[11]

However, the Secretary reported that the reef specimens were so numerous that 'it becomes difficult to prevent tourists from collecting them'. He also reported that, in an attempt to dissuade visitors from taking coral collecting, the caretakers of Lady Musgrave Island, Mr and Mrs Bell, 'discourage the removal of marine growths in every way and to assist in this object specimens of reef life are not even collected for display purposes at the settlement'.[12]

The impacts of coral collectors were not limited to the resort islands, such as Lady Musgrave Island; other islands in the vicinity of resorts were also affected as tourists undertook day-trips. One report, written in 1940 by G. Gentry, a National Parks Ranger, acknowledged that Hoskyn and Fairfax Islands were being damaged since they were visited regularly by tourists from Lady Musgrave and Heron Islands. His report stated: 'There is no doubt that a fair quantity of coral is taken as specimens. Some most outstanding coral beds are to be found around these two islands'. Similarly, E. McKeown, another National Parks Ranger, reported that camping parties from the districts between Cairns and Innisfail that regularly camped on High Island, in the Frankland Group, were removing coral specimens from the Frankland Islands. By October 1940, the foreshore and reef of Green Island had been included on the list of islands from which the removal of coral was prohibited; by the end of the same year, the reefs at Hoskyn and Fairfax Islands, and those at the Frankland Islands, had also been protected.[13]

However, the removal of coral continued. At Green Island, Noel Monkman, the Honorary National Parks Ranger and Honorary Inspector of Fisheries, complained in 1944 about the removal of specimens by American servicemen, stating that:

> I am having an extremely difficult time in protecting the Reef at Green Island from destruction by servicemen spending their weekend leave here. As you are no doubt aware, we have from 200 to 300 men arrive on the island each weekend. I have done my best to prevent the despoiling of the Reef but it is beyond my control. On many occasions when I have requested men to cease breaking off coral and filling their knapsacks with it or collecting kit boxes full of shells and starfish, the men have in many instances become very abusive and aggressive.[14]

Investigation of this issue revealed that Monkman himself, with his brother-in-law, sold corals at the Green Island kiosk (Figure 9.3). In response, Monkman argued that the corals at the kiosk were not taken from Green Island reef; instead, he stated, coral collectors 'have collected these specimens by boat on distant reefs adjacent to the Island, and also purchase from the Island boys [sic] on the luggers visiting Green Island'. After the introduction of legislation, hence, some impacts of collecting were transferred to reefs that were not protected by restrictions.[15]

The damage wrought by coral collectors – including by reef-walking – was apparent to many observers. At Heron Island, Gentry saw 'evidence that shells and coral have been removed in the past'; and, at Green Island, Lock (1955, p207) stated, 'it was evident that some of the coral had been broken apart, and killed, by visitors walking upon it'.[16] Serventy (1955, p77) stated:

> Coral and shell have developed into a minor industry. So much so that most tourist islands in self-defence have had to prohibit the 'picking' of coral and the gathering of shells, at least in large quantities. Boats working from Cairns bring in coral for the tourist trade [...].

138 The impacts of coral and shell collecting

Figure 9.3 Assorted coral displayed at the Green Island kiosk, c. 1940. Source: Uncatalogued photograph obtained from Cairns Historical Society, Cairns, courtesy of G. Jennex

Furthermore, the extent of manipulation of coral reefs had increased to the point where the transplantation of coral from unprotected reefs to those at resort islands, in which coral depletion had taken place, was feasible. By 1952, at Green Island, coral specimens were imported from adjacent reefs in order to supplement the coral gardens that surrounded the underwater observatory, with the result that a total length of 70 feet of coral gardens could be viewed by tourists.[17]

Green Island was not the only location to experience degradation due to coral collecting; Heron Island reef was also depleted by tourist souveniring (Figure 9.4). Commercial coral collecting also took place at Heron Island reef and Wistari Reef, and the depletion of species there was reported by Monkman, when he was the Honorary NP Ranger and Honorary Fisheries Inspector at Heron Island, who stated that:

> The *Don Juan* [...] anchored inside the Heron Island reef for several days, whilst the crew of that boat, *i.e.* two young men and a woman, had been systematically combing the reef during the period of each low tide, both day and night, and had already collected a considerable number of living shellfish and colonies of coral. [...] These people were conducting a business of the

Figure 9.4 Tourists gathering coral specimens from Heron Island reef, c. 1930. Source: QS189/1 Box 17 Item 73, Queensland Industry, Services, Views, People and Events; Photographic Proofs and Negatives; Islands – Barrier Reef, Queensland State Archives, Brisbane

> sale of such specimens by making the shells into jewellery and bleaching and colouring the coral. [...] I went out to this boat, and found coral bleaching on the deck [...].[18]

Monkman also reported that, although that incident took place at Heron Island, the owner of the boat 'did most of his collecting of coral and shells on Wistari Reef, adjacent to Heron Island'.

As a result of the cumulative impacts of tourist souveniring and the increasing impacts of commercial coral collecting, Wistari Reef and One Tree Island reef had deteriorated by 1955. Describing the decline of those reefs, Monkman stated:

> I have been working on the Reef for 25 years as a marine biologist and film producer, and during that period have seen the sad sight of some of our most beautiful reefs being destroyed as thing of beauty and wonder, and have seen the selling of coral and shells become an outrageous racket. Wistari Reef has already commenced to deteriorate through these depredations, and I would implore your Department to protect this reef before it suffers the same fate as

140 The impacts of coral and shell collecting

Figure 9.5 Display of Great Barrier Reef coral for the Qantas office in Tokyo, 1961. Source: QS189/1 Box 17 Item 73, Queensland Industry, Services, Views, People and Events; Photographic Proofs and Negatives; Islands – Barrier Reef, Queensland State Archives, Brisbane

so many other reefs. This also applies to One Tree Reef, but I see no reason at all why all the reefs on the Great Barrier should not be rigidly protected.[19]

Regardless of the prohibition of coral collecting, visitors continued to remove specimens from the Great Barrier Reef throughout the 1960s. Coral was also used for commercial and official purposes; collections were used to decorate Qantas offices (Figure 9.5), for instance, and a large collection, comprising over 1,350 coral specimens and six giant clams was displayed at the 1967 Exposition in Montreal (Peel, 1966, p852).[20]

Other than the informal collecting of souvenirs, coral collecting took place in a more organised manner, encouraged by the Queensland Government, using a system of coral collecting licences. Evidence of those licences survives in the QSA for the period 1962–1969, and nineteen coral areas have been identified using these records, but it is likely that the industry was more extensive than the extant records indicate. The nineteen coral collecting areas that have been reconstructed using archival evidence were located at twelve reefs and islands (Figure 9.6). The distribution of the coral areas indicates that during that period the coral collecting industry exploited reefs in the vicinity of the major ports of Cairns, Townsville, Mackay and Gladstone, with a concentration of activity in

The impacts of coral and shell collecting 141

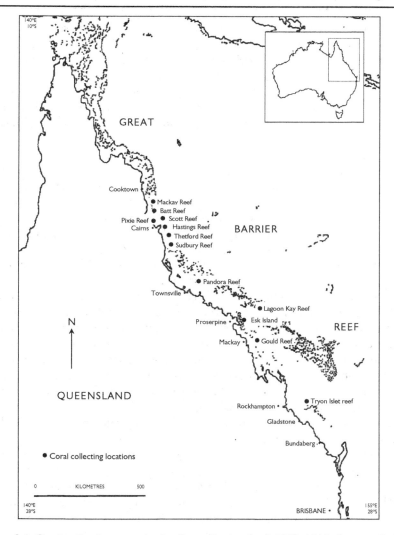

Figure 9.6 Coral collecting areas in the Great Barrier Reef, 1962–1969. Source: Author, compiled from archival files found in Folder 1964, PRV14712/1 Item 788 Box 190, QSA

the Cairns area. Although archival files held at the QSA contain details of the boundaries and lessees of the nineteen coral collecting areas, they do not reveal the criteria by which the coral areas were selected or whether any monitoring of commercial collecting activities occurred.

Some additional evidence provides more detail about the coral collecting industry. Applications for coral collecting licences were made to the Queensland Government and were accompanied by sketch maps of the proposed areas. In addition to the nineteen areas reconstructed using the QSA files, other applications for coral collecting were made, such as A. F. Paterson's application for a licence to remove coral from Otter Reef, near Cardwell. However, regarding that application, the Harbour Master at Townsville stated that:

> Present policy requires that coral leases are normally submerged at all times and remote from public areas. This proposed lease on 'Otter Reef' is a popular fishing ground and anchorage for amateur fishermen. I recommend that this application should be refused.[21]

Therefore, one requirement of the coral areas was that they should not be visible from the surface; the coral areas were required to remain below low-water mark. Nonetheless, some coral was removed from Otter Reef, as Paterson stated that live coral was 'easily obtained at low water and is abundant'.

The impacts of commercial coral collectors were greater than those of individual tourists, although the numbers of the former were far smaller. Commercial operators sometimes took coral from protected reefs, such as Green Island reef, as one oral history informant revealed, and some collectors removed large quantities of corals from individual reefs.[22] An example of commercial coral collecting is shown in Figure 9.7, which illustrates the collecting business of the pioneer aviator, Tom McDonald, who also operated a jewellery trade using coral specimens. McDonald and his co-workers collected coral from reefs in the Cairns area, including Double Island reef, yet no documentary evidence of their business was found in the QSA; their coral collecting pre-dated the coral licence system. Hence extensive commercial coral collecting had probably already occurred by the time that coral leases were first issued.

Therefore, during most of the period of European settlement in Queensland, widespread coral collecting occurred. An indication of the scale of the activity is provided in the Annual Reports of the QDHM, in which the numbers of coral and shell-grit licences issued in Queensland for the period 1931–1968 were reported; by the end of that period, more than 1,000 coral and shell-grit licences had been issued. These data suggest that, by 1962, a significant industry had been established – and probably operated along similar lines – until the formation of the first marine parks in the region, in 1974. As Lawrence et al. (2002, p27) stated:

> Coral collecting remained a popular pastime for tourists. The limited restrictions on collecting under Queensland Fisheries legislation that remained in force well into the 1970s were an indication that coral souveniring continued to be a popular activity. The Queensland Government declared marine park status over two heavily used reef sites, the Heron-Wistari Reef and Green Island Reef in 1974, under the *Queensland Forestry Act 1959*.

Overall, from the earliest regulation of coral collecting in 1933 until the 1980s, coral collection in the Great Barrier Reef increased (Harriott, 2001). Since the

Figure 9.7 Commercial coral collectors working at Double Island reef, c.1930. The individuals shown here are Frank Kelly, Inky Nicholls, Harry Bird and Tom McDonald, aboard the Suva. Jack Clarke was also present. Source: Image No. P09768, Image Library, Cairns Historical Society, Cairns

formation of the GBRMP, the coral fishery has been regulated and collectors now remove around 50 tonnes of material per year from 50 authorised coral areas. The nature of commercial coral collecting has also altered, as Oliver (1985) has acknowledged, from a focus on the souvenir trade – in which one species, *Pocillopora damicornis* ('brown-stem'), dominated the harvest – to supplying the live aquarium industry with high-value species, including soft corals, anemones and other Cnidarians.

Shell collecting in the Great Barrier Reef

Documentary and oral evidence suggests that the cumulative impacts of shell collecting in the Great Barrier Reef have been considerable. In this section, some general impacts of shell collecting are considered, before the more specific damage that has been sustained by giant clams (*Tridacna spp.*) is described subsequently. Shells have attracted the interest of collectors and observers in the Great Barrier Reef since the earliest period of British exploration and settlement in Queensland. Two reports state that James Cook observed 'giant cockles' – giant clams growing 'to a length of ten feet and a weight of a ton' – and ate some of the smaller ones (Bedford, 1928; Christesen, 1936, p31). In October 1844, Joseph Beete Jukes (1871, p234) wrote: 'I then determined to live ashore to arrange my shells'; and in 1872, C. H. Eden (1872, p294) stated:

> There was a beautiful little island called Garden Island, to which the inhabitants of Cardwell used to resort for oyster picnics, and where a great number of cowries of all sizes could be found by turning over the stones at low water.

Bartley (1892, pp218–19) referred to the practice of collecting Australian marine shells, especially different *Conus* species, and he stated that *Cypraea* are 'walked off' from the beaches of eastern Australia. Agassiz (1898, p107), who visited the Great Barrier Reef in 1896, reported that at Stone Island reef, near Bowen, 'the bottom of the bay is covered with fine mud and broken shells', although that damage may have been due to the tropical cyclone that struck Port Denison (now Bowen) on 30 January 1884; further north, Agassiz (1898, p115) wrote: 'We were struck with the great number of dead *Nautilus* and *Spirula* shells thrown up on the sand beaches of the Three and Two Isles groups', and by the abundance of dead cuttlefish bones that were associated with these shells. By the end of the nineteenth century, therefore, these records suggest that European explorers and settlers had made observations of shelled organisms and had collected many specimens.

The impacts of shell collecting intensified during the twentieth century as more collectors worked the reefs and as more locations became accessible to tourists, commercial collectors and shell clubs. The destruction of shell populations and other marine life at Green Island as a result of tourists taking souvenirs, for example, was reported as early as 1929; as tourist resorts developed, the difficulty in preventing increasing numbers of visitors from souveniring shells became apparent.[23] Describing a cruise in the Great Barrier Reef in the early 1930s, Hughes (1937, pp5, 27) reported shell collecting on Langford Island reef flat, and at Redbill Island reef he stated: 'Great slabs of dead coral were overturned and their undersides scanned for the pretty cowrie shells nestling in the crevices'; Redbill Island reef also produced 'a rich harvest' of spider shells. In 1930, Barrett (1930, pp378, 380) wrote: 'Combing the reef for shells is a delightful recreation'.

However, the activities of shell collectors generated opposition as well as enthusiasm; during the 1930s, some individuals advocated regulating the activity and prohibiting shell collecting at some locations. One area about which concern was expressed was the Molle Islands, in the Whitsunday Group, where the lessee, H. G. Lamond, requested the Queensland Government to restrict the taking of shells, for he stated: 'The trouble as I see it – and I have taken particular note – is not what the people take and preserve. It is what they damage in getting specimens'.[24] In 1938, E. O. Marks, the Honorary Secretary of the GBRC, expressed anxiety about 'the harm which must result from promiscuous gathering of marine and other trophies' in the Great Barrier Reef; he acknowledged that the effects of over-collecting were greatest in the most accessible places: those in the vicinity of the tourist resorts.[25]

Yet while Lamond and Marks regarded the activities of collectors as a form of vandalism, other observers were more sympathetic to the actions of tourists. For example, C. J. Trist stated: 'Thoughtlessness rather than vandalism can better describe the desire of this temporary population to souvenir and interfere with the natural beauty of these islands'.[26] However, the effectiveness of the regulation of shell collection often depended on the willingness of caretakers at the island resorts to enforce protective legislation. At Lady Musgrave Island, in 1940, an officer of the QGTB reported the need to curtail shell collecting by tourists, acknowledging that the need to preserve shells on that reef had become pressing.[27]

Exceptions to the restrictions on shell collecting were made for some particular purposes, such as scientific research. For example, researchers at the University of Queensland were permitted to collect shells in the Whitsunday Group (particularly at Hayman Island) on behalf of Professor Goddard during August 1941. A similar period of shell collecting by the Royal Zoological Society, on behalf of the Australian Museum, was permitted at Heron Island in December 1941, and at Green Island in August 1950.[28] The Secretary of the QDHM acknowledged that permits for shell collecting for scientific research represented a special case, and that in general shell collecting should be discouraged. Yet the Queensland Government remained ambivalent about shell collecting, for evidence of over-collecting and damage to reefs in tourist areas accumulated; on the other hand, some shell collections were used to promote both scientific research and the development of the tourist industry in Queensland.[29]

By the 1950s, however, both the extent of shell depletion and the level of concern about damage to the reefs had intensified. The Honorary Secretary of the Caloola Club of Sydney, following a visit to Heron and North West Islands, wrote the following account of the extent of the damage inflicted by shell collectors:

> We were impressed by the amount of poaching and destruction that has and is taking place on the reefs surrounding Heron Island. Numerous shells in which the animals were still alive were seen to be collected by guests, not with the intention of private collection, but for illicit trading: a hat full of live Cone shells, several live Cowries of varying species. On one occasion after a visit to Nor'West [North West] Island, a large live Bailer was collected by a member of the Management Staff and it later came to our notice that lampshades using Bailer Shells were available at a given store for Five Pounds. One member of our party was approached by a guest who had a large quantity of 'very good coral for sale'. That the illicit taking of live material is high is very evident by the depreciation of species since my last visit some ten years ago. Amongst trinkets on sale at the island were large stocks of 'turtles' made from Cowries of two or three species; I do not know the source of supply of these trinkets, but it is certain that it is very difficult to collect

shells of the quality used, without the taking of live material. During a cruise to a neighbouring island, the suspicious behaviour of two craft near the edge of a reef, suggested poaching of coral and associated life, particularly when they quickly weighed anchor and steamed off, out of range.[30]

As a result of that extensive over-collection and poaching of shells, the author concluded that 'we are certain that it has already had repercussions' on the health of the reefs at Heron and North West Islands.

In response, the management of the Heron Island resort acknowledged the cumulative degradation occurring to shell populations. He stated:

> I agree with the submissions of the Secretary of the Club in regard to the gradual depreciations of coastal and marine life, which occurs to a small degree rather constantly, and which [are] a source of continual worry to us. [...] As you know, guests frequently endeavour to take with them a souvenir of their stay, and it is the cumulative effect of this over a period of time which is our major worry. It is possible that certain guests, who would be very few in number, endeavour to obtain material for trading, but I can stress most strongly that this is not countenanced by the Management, and in any cases occurring within our knowledge, we take all action possible to prevent it. The inclination of guests to obtain a souvenir is our principal reason for maintaining a supply of trinkets in the Canteen, but we would point out that the entire supply of items for this purpose is obtained from Cooktown and North Queensland.[31]

Further concerns were expressed to E. M. Hanlon, Premier of Queensland, by the Honorary Secretary of the National Parks Association of Queensland, who acknowledged 'the gravity of the position regarding the Barrier Reef natural resources and wonders', including shell populations, since these resources were being 'stripped bare'. Despite the prohibition of shell collecting, the same author claimed that Mrs Cain, the wife of the Premier of Victoria, had stayed recently at a Great Barrier Reef resort and had brought back a 'marvellous collection of shells'.[32]

The most heavily impacted shell collecting locations were probably the Heron Island, Wistari and Green Island reefs, despite their status as totally protected reefs, because those reefs were the ones most commonly visited by tourists. Subsequently, however, vulnerable areas also included other coral cays within easy reach of the main Queensland ports as regular shelling trips to locations such as Michaelmas and Arlington reefs were organised by shell clubs and were widely advertised (Ladd, 1970, p56). In addition, the depletion of shells was also concentrated in those areas that surrounded protected reefs, but which were easily accessible from the resorts. Julie Booth, who observed the impacts of shell-collectors at Fairfax Island in 1969, stated that visitors from the tourist resort at Lady Musgrave Island 'spend the day here, combing the reef for shells', because shell collecting was prohibited at the reef at Lady Musgrave Island; she reported that one party included more than twenty collectors, who arrived at Fairfax Island from Lady Musgrave Island by boat.[33] Those activities, even if sporadic, probably inflicted significant damage during the shelling visits.

Despite its protected status, Green Island reef continued to suffer degradation from over-collecting. One report of 1958 stated that the reef 'is being stripped by unthinking day trippers and other visitors. At low tide they swarm on the reef with buckets and bags and cart away living coral and shells'.[34] No reduction in that activity was apparent by 1973, when another report stated: 'On a recent visit to Green Island I was appalled to see the number of people returning to the mainland with plastic bags full of coral and shells'.[35] The collection of shells for jewellery manufacture was described as follows:

> [Ron and Mary Rogan] were noted in many parts of the world for the distinctive hand-made jewellery they produced, all of it featuring shells of the Great Barrier Reef. At this time of year the yardman, Jolly McKay, spent some hours each day searching for shells, making a tour of the beaches as the tide went out for specimens washed up, then paddling the reefs in old sandshoes, turning over blocks of coral, thus exposing clusters of tiny living shell-fish.
> (Noonan, 1962, p105)

Further damage from shell collecting at Green Island, and at Michaelmas Cay, were reported in *The Cairns Post* in 1972. Damage also occurred to corals, as a result of shell collectors failing to replace overturned coral boulders; in particular, the 'intense depredation by shell collectors' on fringing reefs to the north of Cairns was reported by Isobel Bennett to be so severe that very little living coral remained on the reefs.[36]

By 1974, degradation due to shell collecting had also been reported at Lizard Island. In that year, describing changes in the Lizard Island reefs, Roger Steene wrote:

> I have been a constant visitor to the Island and its adjacent reefs for the past eighteen years and environmental changes seem to be ever increasing since the establishment of the aerodrome and the discovery of the island as a good anchorage and camping area. During my earlier visits, the *Mauritania* cowrie shell was abundant on the fringing reef on the east side of the island. As time went by and this knowledge became widespread, collectors and others took them until recently, I was not able to find a single specimen. I actually saw a group who had a box with 150 of these shells to sell. Two years ago, I counted seventeen in a half day period.[37]

Also in 1974, Steve Domm, the Director of the Lizard Island Research Station, reported that a charter boat had been at Lizard Island for a week 'with much shell collecting going on, also a small clam had been killed and eaten, plus earlier someone removed a giant clam from the lagoon'.[38] In response to the increasing number of reefs depleted by commercial shell collectors, the Wildlife Preservation Society of Queensland requested increased protection of Lizard Island reefs and the other reefs in the vicinity of Cairns, especially Green Island reef and Michaelmas Reef; over-collection of shells was also reported at the fringing reef at Orpheus Island.[39] Oral history evidence provided by shell collectors generally corroborates the documentary reports of their impact on coral reefs.[40]

Taken together, the evidence presented above indicates that extensive degradation has occurred both to shell populations and to some coral reefs as a result of shell collecting. The long time period during which collecting has taken place in the Great Barrier Reef, the lack of protection of shells during the early period of European settlement in Queensland, the considerable difficulties in enforcing restrictions of shell collecting and the desire of the Queensland government to promote tourism in the Great Barrier Reef have resulted in impacts on shell populations that have been sustained for many decades. Those impacts were concentrated around the major tourist resorts in the Cairns, Whitsunday and Capricorn-Bunker areas: especially at Green Island, Heron Island, Wistari Reef, Lady Musgrave Island and the Lizard Island reefs. To those recreational shell collecting impacts has been added the influence of the commercial shell trade, which also removed large quantities of shells. Therefore, the cumulative impact of shell collecting has probably substantially depleted some locations of their shells, with unknown consequences for the ecology of those reefs.

Impacts on giant clams

Giant clams, *Tridacna spp.* – in particular, *Tridacna gigas* – have attracted the attention of shell collectors since the early period of European settlement. In 1892, the existence of huge *Tridacna* shells in the Great Barrier Reef, 'four of them to a ton', was reported by Bartley (1892, p218). The animals attracted attention for their size and also because of the mistaken belief that they presented a danger to swimmers by trapping their feet, or even swallowing them completely (Thomson, 1966, p38). That perception was articulated in a report of 1935, describing the giant clam population at Low Isles, which also drew attention to the destruction of the animals by visitors:

> At Clam Spit the latest count of the clams that had been rolled there, and that numbered 80 a year ago, now stands at 69. The majority of these animals are favourably situated so that few deaths should normally occur from now on. Unfortunately, visitors to the Island seeing these 'dangerous' animals have a tendency to slash them across, saying as they do, 'You'll never drown another person.' It is an extremely childish action, that results from the continued publishing of the childish story that once upon a time a person put his [sic] foot in a clam which closing on it held him prisoner until the tide came in and drowned him.
>
> (Moorhouse, 1935, p2)

Damage to giant clams was also reported in 1937 at the reef on the southeastern side of Green Island, where visitors habitually took small clams away with them. Clams were also killed *in situ*, as a National Parks Ranger stated: 'Fish spears have on occasions been thrust into clams, killing them'.[41] One example of that type of destruction was given by National Parks Ranger McKeown, who stated that 'some time ago a large clam was brought in from one of the outer reefs, and placed in shallow water for exhibition purposes; recently this clam was speared, and killed'. Consequently, the National Parks Ranger argued, the collection of those animals from Green Island reef should be prohibited.

Giant clams were removed from the reef – or were damaged *in situ* – for a variety of reasons besides popular fear of the danger they presented to swimmers. Barrett (1930, pp378, 380) acknowledged that the demand for unusual shell species – particularly the giant clams – was considerable, and he stated that the valves of *Tridacna gigas* were sought as garden ornaments and home aquaria. Ellis (1936, pp83–4) described the use of giant clams for food, at Raine Island, but also their exploitation as curios, stating that:

> A feature which impressed us considerably at Raine Island was the enormous number of giant clam-shells (*Tridacna gigas*) found in a shallow lagoon, perhaps four feet deep at low tide. [...] An average pair of these enormous bivalves would weigh about three hundred-weight and measure about three feet in length; some indeed were considerably larger. [...] The fish of these *Tridacna* are enormous, but the only portion used by our Chinese labourers for food was the muscle connecting the two sides. It will convey some idea of the size of these gigantic mollusca if it is realized that this muscle usually weighs about five pounds [...]. The giant clams with their inner surface of pure white, like polished marble, are considerably sought after as curios.

An example of the ornamental use of giant clam shells, at Orpheus Island in 1967, is shown in Figure 9.8.

Extensive damage to giant clams in the Great Barrier Reef also occurred as a result of the activities of poachers, particularly from Taiwan, China and Korea. Domm (1970, p44) stated:

> The clam fishery is exploited by Nationalist Chinese and Korean fishermen, and the semi-dried clam meat produced commands an excellent price in the Orient. The adductor muscle is cut from the clams and dried aboard the fishing vessels. The operators generally work in knee-deep water at low tide and, although they leave the clam shells where they are, they damage a large amount of coral wading to them.

Although considerable poaching of *Tridacna gigas* took place after 1970, it is clear that giant clam numbers had already declined since European settlement. Moreover, due to the slow growth rates of those organisms, the overall increased

Figure 9.8 Giant clam shells at Orpheus Island, 1967. Source: SRS189/1 Box 17 Item 73, Queensland State Archives, Brisbane

mortality of giant clams caused by tourists, clam fishers and poachers is likely to have been unsustainable.

However, the effects of tourists, clam fishers and poachers did not represent the only impacts on giant clams. Despite official protection of many marine species, the Commonwealth Government arranged in 1966 for the removal of giant clams from the Great Barrier Reef; these specimens probably formed part of the coral reef display in the 1967 Exposition at Montreal. Peel (1966, p852) stated:

> In April, 1966, a clam-collecting party working from and with the assistance of the *Cape Moreton* (Commonwealth Department of Shipping and Transport) obtained six unusually large specimens of giant clams for incorporation in the [coral reef] display.

The removal of six unusually large giant clams from the Great Barrier Reef at the request of the Commonwealth Government as recently as 1966 suggests that limited conservation of that organism was encouraged, even by government officials, before the creation of the GBRMP in 1975.

Summary

The evidence presented in this chapter indicates that coral and shell collecting has occurred in many parts of the Great Barrier Reef throughout the period since European settlement, and that particular degradation of some coral reefs (including those at Heron, Hayman, Green, Masthead and Lady Musgrave Islands, and at Wistari Reef) has occurred. That degradation was due to the combined impacts of prolonged, cumulative coral and shell souveniring by tourists, and to the removal of large amounts of material by commercial collectors. Coral collecting resulted in cumulative impacts which were reported to have been severe at the major tourist resorts of the Great Barrier Reef, including those resorts at Heron, Green, Lady Musgrave and Lizard Islands. Those impacts were sustained over more than fifty years and were concentrated in the most accessible parts of the most frequently visited reefs. Despite restrictions of the removal of coral since 1932, both licensed and unlicensed collecting continued; documentary and oral history sources describe the deterioration of coral reefs that accompanied coral collecting. In addition to the souveniring of coral by many visitors, the coral collecting industry – regulated by the Queensland Government – was responsible for the removal of large amounts of the most attractive coral from at least twelve coral reefs. As a result, coral collecting has contributed to the decline of many coral reefs, particularly nearshore and fringing reefs.

The impacts of shell collecting in the Great Barrier Reef have also been widespread, prolonged and cumulative. Some evidence suggests that a reduction in the average size of some species has taken place since extensive recreational shell collecting commenced, indicating that some species have been collected excessively. Particular depletion of shell populations was reported at Green, Heron, Lady Musgrave and Fairfax Islands, at Wistari Reef and at Michaelmas Cay, although other reefs that are easily accessible from the Queensland coast were also exploited by shell collectors, including Yule Point Reef. It is clear that some shell collectors operated without regard for the sustainability of shell populations, for the damage caused to coral reefs while obtaining specimens, or for prohibitions on the removal of marine species from protected reefs. Particular impacts were sustained by giant clams (*Tridacna gigas*) as a result of clam fishing, the use of giant clam shells as curios and ornaments, and the destruction of the animals by divers and bathers. As giant clams are long-lived, slow-growing organisms, the giant clam populations in the GBRWHA have probably been considerably depleted since European settlement. However, some rehabilitation of those populations has been attempted as a result of successful aquaculture of giant clams. Overall, therefore, coral and shell collecting represent significant – yet previously neglected – causes of environmental change in many reefs of the Great Barrier Reef.

Notes

1. OHC 13, 4 August 2003.
2. *Order in Council*, 1 June 1933.
3. OHC 31, 4 October 2003.
4. OHC 4, 14 January 2003.
5. See the statistics provided in SCQ, 1879, p174.
6. Out-letter Ref. 32/3263, Secretary, Provisional Administration Board, QDHM, Brisbane to Under-Secretary, Treasury, Brisbane, 10 May 1933, PRV8340/1 Item 1, QSA.

152 The impacts of coral and shell collecting

7   In-letter Ref. 06598, H. G. Lamond, Molle Islands to Under-Secretary, QDAS, Brisbane, 15 March 1933, PRV8340/1 Item 1, QSA; In-letter Ref. 4488, H. G. Lamond, Molle Islands to QDHM, Brisbane, 10 August 1933, PRV8340/1 Item 1, QSA; *Order in Council*, 1 June 1933.
8   *Order in Council*, 27 September 1937, *Order in Council*, 20 July 1939.
9   In-letter Ref. 38/14394, E. O. Marks, Honorary Secretary, GBRC, Brisbane to Hon. F. A. Cooper, Treasurer, Brisbane, 12 December 1938, PRV8340/1 Item 1, QSA.
10 Mr A. W. Bauer, cited in In-letter Ref. 39/6316, Director, QGTB, Brisbane to Secretary, QDHM, 1 June 1939, PRV8340/1 Item 1, QSA; *Order in Council*, 20 July 1939.
11 In-letter Ref. L40.2373.11, Secretary, Office of the Commissioner for Railways, Brisbane to Secretary, QDHM, Brisbane, 22 May 1940, PRV8340/1 Item 1, QSA.
12 In-letter Ref. L40.2373.11, Secretary, Office of the Commissioner for Railways, Brisbane to Secretary, QDHM, Brisbane, 22 May 1940, PRV8340/1 Item 1, QSA.
13 In-letter Ref. 225/45, Geo Gentry to Secretary, Sub-Department of Forestry, Brisbane, 11 October 1940, PRV8340/1 Item 1, QSA; In-letter Ref. 225/45, E. McKeown, National Parks Ranger, Tully to Secretary, Sub-Department of Forestry, Brisbane, 20 September 1940, PRV8340/1 Item 1, QSA; Out-letter Ref. 4868, Secretary, QDHM, Brisbane to Secretary, Sub-Department of Forestry, Brisbane, 29 October 1940, PRV8340/1 Item 1, QSA; Extract 40/13764, G.D.Q. 225/45, J. D. W. Dick, Chief Inspector of Fisheries, QDHM, Brisbane to Secretary, Forestry Sub-Department, Brisbane, 4 December 1940, SRS5416/1 Box 10 Item 59, NP219 Bunker, QSA; Extract 40/13764, G.D.Q., 225/45, J. D. W. Dick, Chief Inspector of Fisheries, QDHM, Brisbane to Secretary, Forestry Sub-Department, Brisbane, 4 December 1940, SRS5416/1 Box 10 Item 58, NP220 Bunker, QSA.
14 In-letter Ref. JRD/LL, Noel Monkman, Green Island Kiosk to E. McKeown, Forestry Officer, Tully, 1 June 1944, SRS5416/1 Box 66 Item 447, NP836 Trinity 'R' – Green Island – Protection of marine life, QSA, p4.
15 In-letter, Honorary Secretary, NQNC, Cairns to Mr Jones, Minister for Lands, Brisbane, 17 July 1944, SRS5416/1 Box 66 Item 447, NP836 Trinity 'R' – Green Island – Protection of marine life, QSA; Out-letter, J. D. W. Dick, Chief Inspector of Fisheries, Brisbane to Secretary, Queensland Forestry Sub-Department, Brisbane, 23 February 1945 SRS5416/1 Box 65 Item 443, NP836 Trinity 'J' – Green Island, QSA; Noel Monkman, 'Copy of Letter Attached', 25 January 1945, SRS5416/1 Box 65 Item 443, NP836 Trinity 'J' – Green Island, QSA.
16 Letter written by G. Gentry, National Parks Ranger, 10 May 1944, SRS5416/1 Box 10 Item 61, NP231, Bunker – Heron Island, QSA.
17 In-letter, C. J. Trist to National Parks Ranger McKeown, Tully, 17 January 1952, R836 Trinity 'P', QSA.
18 In-letter Ref. 2A/D0, 152/4335, Noel Monkman, Honorary Ranger, Honorary Fisheries Inspector, Heron Island to W. Wilken, Secretary, Brisbane, 3 January 1955, SRS5416/1 Box 66 Item 446, NP836 Trinity 'P' – Green Island – Underwater observation chamber, QSA, pp1–2, p1.
19 In-letter Ref. 2A/D0, 152/4335, Noel Monkman, Honorary Ranger, Honorary Fisheries Inspector, Heron Island to W. Wilken, Secretary, Brisbane, 3 January 1955, SRS5416/1 Box 66 Item 446, NP836 Trinity 'P' – Green Island – Underwater observation chamber, QSA, pp1–2, p1.
20 QS189/1 Box 17 Item 73, Queensland Industry, Services, Views, People and Events; Photographic Proofs and Negatives; Islands – Barrier Reef, QSA.
21 In-letter, Harbour Master, QDHM, Townsville to Secretary, QDHM, Brisbane, 3 July 1964, Folder 1964, PRV14712/1 Item 788 Box 190, QSA; In-letter, A. F. Paterson,

Southport to QDHM, 1 June 1964, Folder 1964, PRV14712/1 Box 190 Item 788, Subject batches – Oyster, coral and shell grit, QSA.
22. OHC 31, 4 October 2003.
23. Out-letter, Town Clerk, Cairns to Under-Secretary, Treasury, Brisbane, 12 December 1929, SRS146/1 Item 2, QSA; Out-letter, Town Clerk, Cairns to Chairman, Queensland Marine Board Office, Brisbane, 26 September 1931, SRS146/1 Item 2, QSA.
24. In-letter Ref. 06598, H. G. Lamond to Under-Secretary; In-letter Ref. 4488, H. G. Lamond to QDHM, 15 March 1933, PRV8340/1 Item 1, QSA.
25. In-letter, 38/14394, E. O. Marks, Hon. Secretary, GBRC, Brisbane to the Hon. F. A. Cooper, Treasurer, Treasury, Brisbane, 12 January 1938, PRV8340/1 Item 1, QSA.
26. Circular No. 727, C. J. Trist, Secretary, Queensland Sub-Department of Forestry, Brisbane, 'Memo: Protection of Islands – Barrier Reef', 23 March 1939, PRV8340/1, Item 1, QSA.
27. In-letter, Secretary, Office of the Commissioner for Railways, Brisbane to Secretary, QDHM, Brisbane, 22 May 1940, PRV8340/1 Item 1, QSA.
28. Out-letter, F. A. Cooper, Treasurer, Queensland Treasury, Brisbane to Miss M. Cross and Miss P. Hardy, University of Queensland, Brisbane, 14 August 1941, PRV8340/1 Item 1, QSA; Out-letter Ref. 41/10354, Secretary, QDHM, Brisbane to Miss G. Thornley, Lidcombe, New South Wales, 15 December 1941, PRV8340/1 Item 1, QSA; Out-letter, Chief Administrative Officer and Secretary, QDHM, Brisbane to Secretary, Queensland Sub-Department of Forestry, Brisbane, 5 January 1950, SRS5416/1 Box 66 Item 447, NP836, Trinity 'R' – Green Island – Protection of Marine Life, QSA.
29. In-letter, Director, Queensland Tourist Services, Brisbane to Under-Secretary, Queensland Department of Mines and Immigration, Brisbane, 20 August 1954, RSI920/1 Item 8, General correspondence batches – Previous files, QSA.
30. In-letter, Director, QGTB, Brisbane to Secretary, QDHM, Brisbane, 21 March 1955, RSI920/1 Item 9, General correspondence batches – General Tourist Bureau matters, QSA, p1.
31. In-letter, Director, QGTB, Brisbane to Secretary, QDHM, Brisbane, 21 March 1955, RSI920/1 Item 9, General correspondence batches – General Tourist Bureau matters, QSA, p2.
32. In-letter, J. K. Jarrott, Honorary Secretary, National Parks Association of Queensland, Brisbane to the Hon. E. M. Hanlon, Premier of Queensland, Brisbane, 3 October 1947, RSI920/1 Item 9, QSA, pp1–2.
33. In-letter, J. Booth, Fairfax Island to W. Wilkes, Secretary, 30 September 1969, SRS5416/1 Box 10 Item 58, NP220, Bunker, QSA.
34. Article, 'Hands off Green Island', *Courier-Mail*, 20 September 1958, SRS5416/1 Box 66 Item 447, NP836, Trinity 'R', QSA.
35. Anne Taylor, Darlinghurst, New South Wales to Queensland Director of Forestry, Brisbane, 12 October 1973, SRS5416/1 Box 66 Item 447, NP836, Trinity 'R', QSA.
36. Article, 'Vandalism at Green Island?', *The Cairns Post*, 23 November 1972, NP836 Trinity 'A', Green Island, QSA; I. Bennett, 'Audio-visual: the Great Barrier Reef', Manuscripts, Box 6, Folder 18, Miscellaneous, 1967–1995, MS9348, Papers of Isobel Bennett, 1944–2000, NLA, pp12, 23.
37. R. Steene, Cairns to S. Domm, Director, Museum Research Station, Lizard Island, 20 May 1974, SRS5416/1 Box 28 Item 179, NP153, Flattery 'A' – Lizard Island, QSA, pp1–2.
38. In-letter, S. Domm, Resident Director, Lizard Island Research Station to 'Alan', 3 August 1974, SRS5416/1 Box 28 Item 179, NP153, Flattery 'A' Lizard Island, QSA.

39 In-letter, Joan M. Wright, President, Wildlife Preservation Society of Queensland Inc. to S. Domm, Resident Director, Lizard Island Research Station, 22 July 1974, SRS5416/1 Box 28 Item 179, NP153, Flattery 'A' Lizard Island, QSA.
40 Anonymous, 'Recollections of the reef', Changes in the Great Barrier Reef since European Settlement, Oral History Collection, School of TESAG, JCU, September 2003; M. Ford, 'Shell populations in the Capricorn-Bunker group of the Great Barrier Reef', November 2003, Changes in the Great Barrier Reef since European Settlement, Oral History Collection, School of TESAG, JCU, 3pp.
41 In-letter, E. McKeown, National Parks Ranger, Tully to Secretary, Brisbane, 10 April 1937, SRS5416/1 Box 66 Item 448, NP836, Trinity 'R', QSA.

Chapter 10

The impacts of guano and rock phosphate mining

Introduction

More than 300 coral cays and 600 continental islands lie within the GBRWHA and they form distinctive environments of the Great Barrier Reef. The evolution and geomorphology of cays and continental islands in the Great Barrier Reef have been discussed by Hopley (1982), who acknowledged that some of those cays and islands have been significantly modified by human activity since European settlement. Indeed, the transformation of some island landscapes probably represents the most comprehensive human impact on the landscape of the Great Barrier Reef, at least at the local scale (Flood, 1977, 1984; Stoddart et al., 1978). Examples that illustrate the extent of that transformation, and which are discussed in this chapter, include the changes wrought by guano miners at Lady Elliot Islands, Raine and North West Islands, and the removal of rock phosphate from Holbourne Island. First, however, the earliest recorded European impact on an island in the Great Barrier Reef is described: the construction of the navigation beacon at Raine Island, in 1844, using phosphatic sandstone quarried at the island. Next, the various guano and rock phosphate mining operations in the Great Barrier Reef are considered, together with their impacts on island landscapes.

The construction of the navigation beacon at Raine Island, 1844

The earliest significant construction on a Great Barrier Reef island occurred at Raine Island in 1844, when a navigation beacon was built to assist ships sailing near the hazardous reefs in the locality. A beacon was required in the northern part of the Great Barrier Reef to mark the entrance to the Blackwood Channel, through which ships could pass relatively safely to the Great Barrier Reef lagoon; the construction of a feature that was visible from twenty nautical miles away meant that navigators did not have to approach the outer reefs until they had established their position. Joseph Beete Jukes (1847, p266), during his visit to Raine Island, sketched the cay, the newly-constructed beacon and the temporary settlement present at that time. The beacon tower also served another purpose: it was stocked with provisions for shipwrecked mariners. The construction of the beacon commenced in May 1844 and the work was carried out by a convict labour force (Lawrence and Cornelius, 1993). By September of that year, the beacon was completed.

The beacon was built using phosphatic sandstone blocks that were quarried from the eastern part of the island, and lime that was obtained by burning *Tridacna* and *Hippopus* shells; Jukes (1847, p266) wrote that the latter were 'to be got in abundance from the reef at low water'. Timber was taken from the wreck of the *Martha Ridgeway*, as was the ship's tank, which was used to collect rainwater (Loch, 1984). The completed tower comprised a circular tower, 45 feet in height and 30 feet in diameter at its base. The walls were five feet thick, and a domed roof carrying a large ball raised the total height of the structure to 63 feet. The large size of this structure, on a relatively small island (approximately 32 hectares), indicates that Raine Island sustained a significant geomorphological impact as a result of the quarrying of the phosphatic rock, as Hopley has acknowledged (1982, p337). In addition, the removal of *Tridacna spp.* and *Hippopus spp.* must have occurred on a considerable scale and caused localised depletion of those species. However, those impacts were overwritten by the more extensive alteration of the island that took place from 1890–1892 as a result of guano mining.

Guano and rock phosphate mining, 1860–1940

Guano -- the cemented deposits formed by accumulations of bird droppings -- and rock phosphate represent natural resources that have been extracted from some islands of the Great Barrier Reef in order to supply phosphatic fertiliser for agriculture. The mining of guano and rock phosphate in the Great Barrier Reef has resulted in many changes in islands, ranging from minor modification of vegetation to the alteration of the geomorphology of entire islands (Heatwole, 1984; Hopley, 1988, 1989). At least ten locations in the Great Barrier Reef have been mined for guano and rock phosphate; those locations are shown in Figure 10.1. A variety of practices occurred in the guano and rock phosphate mining industries: some cays, such as Raine Island, were mined intensively with rapid depletion of the commercial resources. Others, such as Upolu and Michaelmas Cays, were used less intensively, but over much longer periods. Therefore, this section describes a group of diverse mining practices, locations and historical periods, based on evidence found in historical books and in the archival files of the QEPA, held at the QSA.

The date of the earliest guano mining in the Great Barrier Reef is disputed; one account, by Golding (1979, pp77–8), stated that the industry was pioneered by William L. Crowther, of Hobart, who applied to the New South Wales Government for licences to mine guano from Wreck Reef and Cato's Bank in 1861. Golding (1979) claimed that, before the permits were issued, Crowther had commenced removing guano from Wreck Reef; one hundred tons of guano

The impacts of guano and rock phosphate mining 157

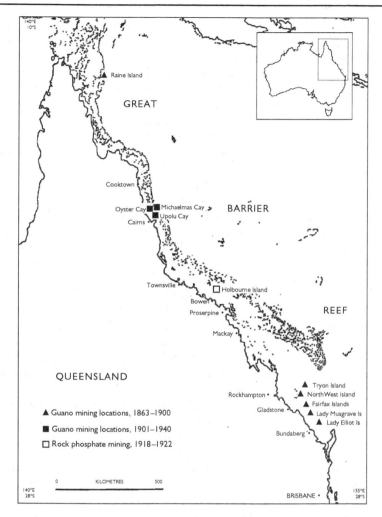

Figure 10.1 Guano and rock phosphate mining locations in the Great Barrier Reef. Source: Author

had been loaded onto the *Harp* when that boat was shipwrecked on the reef. Informal mining of this type probably took place in the period before the 1860s, when the industry became established, and also in subsequent decades, because control of the industry was hindered by political disputes about the jurisdiction of the offshore islands in the Great Barrier Reef (Foxton, 1898; Cumbrae-Stewart, 1930; Golding, 1979).

The first instance of licensed guano mining took place at Lady Elliot Island, from 1863–1873. In 1863, a tender to mine guano from the island by Mr J. Askunas was granted, at a cost of £300 per year; in 1864, Askunas transferred his lease to Crowther, who continued the operation (Walsh, 1987, pp30–2; QNPWS, 1999a, p5). After its extraction from the island, the guano was dried, broken down and collected into sheds before being loaded onto barges, and a system of tramways, sheds and moorings for the barges was constructed. The impact of guano mining on the island was severe; a layer more than 2.5 metres thick was removed from the surface of the island. More than a century later, in 1971, the Secretary of the GBRC stated that the vegetation of the island had still not recovered from intensive disturbance as a result of guano mining, and up to half the area of the cay may have been eroded following the removal of the guano layer.[1] Similarly, Hopley (1982, p340) wrote that little of the original vegetation remained. Heatwole (1984, p39) also found that the environment of Lady Elliot Island had been significantly disturbed by guano mining: most of the vegetation and surface material had been removed by the industry and 'old diggings, tramways, washing mounds and wells' were still detectable. Indeed, Heatwole (1984, p41) concluded, 'Lady Elliot Island's prime ecological value is as a reminder of how destructive uncontrolled human activities can be to a coral cay, and of how prolonged those effects can be'.

After the operation at Lady Elliot Island ceased in 1873, a break apparently occurred in guano mining. However, a decade of further, intensive guano mining took place from 1890–1900. During that period, several other islands were mined in the Capricorn-Bunker Group, and mining also commenced at Raine Island. Previously, during his visit to Raine Island in 1844, Jukes (1847, p266) had commented on the enormous numbers of birds on the island, which produced 'a vast deposit of guano little inferior in quality and value to the famous Peruvian variety'. In 1865, Crowther was issued with a licence to remove guano from Raine Island for seven years; his lease was subsequently transferred to the Anglo-Australian Guano Company and it was renewed in 1871 (Golding, 1979, pp80, 82). Some uncertainty exists about whether guano mining occurred at Raine Island during the period between the issue of that licence and the commercial operation that commenced in 1890; Loch (1984, p183) claimed that, although leases for guano mining were granted for Raine Island as early as 1862, the island was not mined at that time because of doubts about the commercial viability of the guano deposits.

While the activities of the earlier period are uncertain, many documentary sources indicate that intensive guano mining took place at Raine Island from 1890–1892. The operations, carried out by J. T. Arundel and Company, under the management of Albert Ellis, employed a large indentured labour force – of approximately 100 Chinese and Malay workers – and ten European supervisors (Lawrence and Cornelius, 1993, p5). The huts, tramway, locomotive and jetty were installed at that time in order to transport 'tens of thousands of tons of phosphate' from the island to the ships (Hopley, 1988, pp34–5; Loch, 1994, p183). By 1892, however, the mining ceased and the huts, tramway and jetty were removed from the island. A depression, created by the open cast mining, remained in the centre of the island, which is still visible and appears on modern maps (Limpus, 1978, p213). Hopley (1982, pp335, 337) regarded the damage to Raine Island as probably the most devastating impact on any of the islands of the outer Great Barrier Reef; the island was 'completely altered' by the removal of the guano.

In addition to the operation at Raine Island, guano mining took place in the Capricorn-Bunker Group of islands, where profound impacts on vegetation were sustained as a result of the industry. During the 1890s, guano mining was carried out at Fairfax, North West and Lady Musgrave Islands, and evidence of the mining remains in the landscape of those islands (QNPWS, 1999a, pp5–6).[2] In particular, extensive guano mining occurred at Fairfax Islands (which form a double island); one report by a National Parks Ranger, written in 1936, stated that the island 'has been worked very extensively many years ago and large quantities of guano have been removed'.[3] The National Parks Ranger also stated that, by 1936, all of the commercially viable guano had been removed, and he reported that the mining had extended over almost the entire island and only a few acres in the centre of the island remained undisturbed: this central part formed the only section of the island where any vegetation remained, which consisted only of *Pisonia umbellifera*.

One account of guano mining at Fairfax, North West and Lady Musgrave Islands was provided by Ellis (1936, p162), who stated that:

> Fairfax Island was a difficult place to work [...]. The phosphate guano too was much mixed with immense quantities of coral slabs and shingle; the large piles of this material left on the island are good evidences as to the amount of labour we put in. Operations on a minor scale were carried on at Lady Musgrave at the same time, a ketch being employed to lighter cargoes across to the sailing vessels loading at the other island. [...]. A prospecting trip round the Capricorn Group was carried out on the cutter *Lorna Doone* during 1898. Deposits of medium quality were found on North-West Island. These were worked when Fairfax was finished.

Of those three islands, particular degradation occurred at North West Island, which was mined from 1898–1900, as the QNPWS (1999a, p5) acknowledged. Golding (1979, p90) reported that the labour force comprised 107 Asian workers and five Europeans, and the infrastructure included a tramway that was laid across the island and a jetty that was built to the edge of the reef (see also *Brisbane Courier*, 9 February 1900; Lawrence et al., 2002, p20). In November 1899, 550 tons of guano were shipped on the *Van Royal* and another boat – the *Silas* – carried 1,100 tons from the island. Golding (1979, pp90–1) stated that, by February 1900, a total of 4,146 tons of guano had been removed from the island: most of that was exported to New Zealand. During his visit to North West Island in 1936, Steers (1938, p65) observed 'noticeable erosion' of the cay that had been exacerbated, he suggested, by the removal of guano. The impacts of guano mining caused Heatwole (1984, p28) to describe North West Island as 'the most disturbed of the uninhabited islands' in the Capricorn-Bunker Group.

Of the other islands in the Capricorn-Bunker Group, Lady Musgrave Island was worked by guano miners during the 1890s, but little is known about the scale of that operation. However, ridges on the island resulting from the removal of guano were visible to Steers (1938, p54) during his visit in 1936. Tryon Island was probably mined for guano from 1898–1900, but Heatwole (1984, p28) suggested that those operations must have been small, since few indications of mining remain in the landscape. The National Parks Ranger who visited Hoskyn Island in 1936 reported that 'only a few tons of low grade guano occurred' and, probably, neither of the Hoskyn Islands were mined for guano; in contrast to the higher-grade guano deposits worked at Lady Elliot and North West Islands, the extraction of material from the Hoskyn Islands was not economically viable.[4] By 1900, the most intensive guano mining had ceased, in Hopley's (1989, p20) view, because the commercial resources had been rapidly exhausted.

However, guano mining continued after that year at Michaelmas, Oyster and Upolu Cays, near Cairns, in a less intensive manner, but for longer periods (QNPWS, 1998b). In 1901, Captain Robertson was granted a 21-year lease by the Queensland Government to mine Oyster Cay, 'on which there is a large deposit of guano'. One report claimed that, over the period of his lease, Robertson removed 'over a thousand tons of deposit' from Oyster and Upolu Cays (*The Cairns Morning Post*, 28 May 1901, p2; Loch, 1991, p5). It is unclear whether or not that guano was used to fertilise sugar cane fields on the adjacent Queensland coast; however, exports of the product were recorded, for example, to Japan.[5] The operation raised some public concerns about the destruction of the cays. One individual wrote to the Queensland Minister for Mines, asking, 'Could you do anything to prevent Upola Bank [sic] and Oyster Cay on the Barrier Reef being destroyed by removing the coral and guano from these banks?'[6] Nevertheless, the mineral leases for those cays were renewed in 1922 and the removal of guano continued (Loch, 1991, p5).

By 1918, an alternative source of phosphate to the guano obtained from the cays of the Great Barrier Reef had been discovered: the rock phosphate deposits found on Holbourne Island, near Bowen.[7] During the First World War, superphosphate for agricultural fertiliser was sold in Queensland at a cost of £8 per ton; some investors considered that the Holbourne Island material might form a cheaper source of phosphate. The Holbourne Island Phosphate Co. Ltd was formed to investigate and work the deposits for the Australian and New Zealand markets, and agricultural fertilisers were subsequently produced and sold.[8] The company took over Holbourne Island Guano Licence No. 1, which was previously held by Messrs A. Junner and W. M. Gall; that lease was reissued as Mineral Leases Nos. 66 and 67, which was then replaced by Mineral Lease No. 73. An initial geological survey suggested that around 400,000 tons of phosphate were found on the island; a settlement was then constructed on the island, a tramline was laid, the quarried material was carried to the beach using horses and the phosphate was transported to barges using punts (Saint-Smith, 1919, pp122–4).

Phosphate mining at Holbourne Island commenced in 1918, but the grade of the phosphate was found to be too low to supply the inter-state and international markets profitably.[9] However, the material was suitable for local markets, and phosphate was transported from the island via Bowen to Brisbane and Townsville for processing. One source recorded the import of 25 tons 4 cwt of phosphate to Bowen Harbour by the A.U.S.N. Co. Ltd in May 1918.[10] A total of 450 tons of material was removed from the island in 1918; in 1919, the amount increased to 850 tons. However, in 1920 the annual yield declined to 450 tons, and in 1921 only 369 tons 10 cwt (valued at £1,570) were shipped (Linedale, 1922, p600; *The Bowen Independent*, 29 January 1971, p4). Hence, the industry was short-lived and the company ceased its operations at the end of 1921. The following factors contributed to the decline of the industry:

(a) high production costs due to unreliable shipping;
(b) the high cost of freight to the mainland;
(c) the lack of drying facilities on the island, increasing the weight of the shipments;
(d) labour and provisioning difficulties;
(e) the low tonnage output; and
(f) a high proportion of lime contained in the Holbourne Island phosphate, which made the cost of manufacturing superphosphate too high, in comparison with other sources.

(*The Bowen Independent*, 29 January 1971, p4)

After 1921, no further working of phosphate took place at Holbourne Island; subsequently, phosphate was imported from Nauru and Ocean Islands instead and, in the 1970s, the Holbourne Island deposits were declared not commercially viable (QEPA, 2003b).[11] Subsequently, Hopley (1982, p376) confirmed that evidence of the rock phosphate quarry remained in the landscape of Holbourne Island.

Summary

Profound changes in the landscapes of some of the Great Barrier Reef's islands and cays, spanning almost the entire period of European settlement in the region, have been described in this chapter. Island landscapes were modified by the construction of the navigation beacon at Raine Island, in 1844, and as a result of guano and rock phosphate mining at various other islands and cays. Overall, the impacts sustained in the Great Barrier Reef as a result of guano and rock phosphate mining were relatively widespread – occurring in at least ten locations – and prolonged: from 1860 until around 1940. However, those impacts also varied in their intensity as different mining strategies were adopted, and as deposits of varying qualities were worked. The earliest instances of guano mining in the Great Barrier Reef were informal and unlicensed; however, by the 1860s, the industry was organised using a system of guano licences and it attracted considerable capital investment. The guano at Lady Elliot and Raine Islands was stripped rapidly; by 1900, guano mining had also taken place on other islands in the Capricorn-Bunker group, with significant degradation occurring at North West Island. However, not all of the islands that were mined contained such quantities of guano, nor experienced such devastation. From 1901, the pattern of guano mining changed; over several decades, Michaelmas, Oyster and Upolu Cays were mined and thousands of tons of guano were removed. In contrast, the more intensive attempt to remove rock phosphate from Holbourne Island proved too costly to sustain; nevertheless, more than one thousand tons of material were taken.

Some severe geomorphological and ecological transformations accompanied guano and rock phosphate mining in the Great Barrier Reef. In addition to the profound modification of the topography and vegetation of Lady Elliot, Raine and North West Islands, other unintentional changes occurred as a result of the activities of the miners. For instance, Bedford (1928) reported that the descendants of domesticated fowls were found on islands that had been worked for guano, because chickens were kept by the miners as a source of food. Another ecological impact has been sustained by marine turtles at Raine Island: when female green turtles come ashore to lay their eggs at that important rookery, they may encounter hazardous cliffs in the landscape of the cay that were created by the historical mining and quarrying there. Hopley (1989, pp19–20) argued that, at those islands where the geomorphological impacts of guano mining have been severe – especially Raine, Lady Elliot and North West Islands – their recovery may take hundreds of years, if in fact those impacts are reversible. The account of guano and rock phosphate mining presented in this chapter therefore illustrates the variable nature of the early industrial use of the islands and cays of the Great Barrier Reef, as well as the diverse impacts, rates of recovery and effects of those activities. Although only a small proportion of the cays and islands in the Great Barrier Reef were affected by guano and rock phosphate mining, those impacts have nevertheless been significant at the scale of the islands and cays affected.

Notes

1. Mather, 'Statement', SRS5416/1 Box 10 Item 60, NP268, Bunker, QSA, pp1, 5.
2. OHC 44, 4 December 2003.
3. In-letter Ref. 225/2, National Parks Ranger to Secretary, 6 October 1936, SRS5416/1 Box 9 Item. 57, NP224, Bunker – Lady Musgrave Island, QSA, p2.
4. In-letter Ref. 225/2, National Parks Ranger to Secretary, 6 October 1936, SRS5416/1 Box 9 Item. 57, NP224, Bunker – Lady Musgrave Island, QSA, p3.
5. See, for example, the statistics provided in *SSQ*, 1920, p111.
6. In-letter, Mr F. H. Dean, Kuranda to Mr Atherton, Minister for Mines, Queensland Department of Mines, Brisbane, 20 November 1931, PRV8340/1 Item 1, QSA.
7. This information was obtained from an undated edition of 1941 of *The Bowen Independent*, reprinted in *The Bowen Independent*, 29 January 1971, p4.

8 This information was obtained from *The Australian Sugar Journal*, 5 August 1921, p273, courtesy of Dr P. Griggs.
9 'Shipment of phosphates', *The Bowen Independent*, 22 February 1919.
10 Harbour Board, Bowen, Statistical Book No. 1, July 1915 – February 1926, RSI5551/1 Item 1, Statistical books, QSA.
11 In-letter, A.C.F. and Shirleys Fertilizers Ltd to Secretary, Cairns Harbour Board, Cairns, 19 July 1950, RSI13111/1 Item 84, Batches, Harbour Board – Cairns – Leases to ACF and Shirleys, QSA, p2.

Chapter 11

The impacts of coral mining

Introduction

Coral mining has had significant impacts on some parts of the Great Barrier Reef, yet the extent and severity of that activity have not been widely appreciated, despite the fact that public opposition to a proposal in 1967 to mine coral from Ellison Reef, near Innisfail, catalysed significant environmental protests in Australia and was one factor leading to the formation of the GBRMP (Carruthers, 1969, p47; Hopley, 1988, pp34–5; 1989, p20; Bowen and Bowen, 2002, p291). This chapter focuses on the impact of the coral mining industry, which operated in the Great Barrier Reef between at least 1900 and 1940. That industry has been previously neglected in histories of the ecosystem, but it was responsible for removing thousands of tons of coral from some reefs and pulverising it to produce agricultural and industrial lime, which in turn was used on sugar cane fields and in sugar refining mills in coastal Queensland. On some reefs, coral mining was both sustained and intensive, and significant destruction of those reefs must have occurred; coral mining has also affected the landscapes of some of the islands and cays in the region. Twelve locations at which coral mining occurred have been identified; some, such as Snapper Island reef (near Mossman) and Kings Reef (near Innisfail), sustained severe damage as a result of the use of gelignite or crowbars to remove coral.

The coral mining industry was promoted by the Queensland Government and was organised using a system of coral licences, but evidence of additional, unlicensed coral mining was also found. Information about coral mining was obtained from many records of the QDHM and the QEPA. In particular, material was found in the files relating to the preservation of coral from exploitation, the issue of coral licences and the *Fish and Oyster Acts, 1914–1935*. However, the records held at the QSA begin and end abruptly, with discontinuities between series; archivists at the QSA suggested that other files may have been lost when the Departmental offices in Brisbane were inundated during the Australia Day floods of 27 January 1974. The sequence of coral licences suggests that more areas may have been mined for coral than those specified in the surviving records. Documentary evidence also suggests that unlicensed coral mining occurred in some places, such as Kings Reef, before the system of coral licences was introduced.

Furthermore, oral history informants revealed that coral mining occurred at Snapper Island – a location for which no evidence of a coral licence was found. Therefore, coral mining may have occurred more extensively than this account indicates. Oral sources also provided additional details of the process of coral mining, the infrastructure used in the industry and the impacts that remain in the landscape.[1] Although it is difficult to reconstruct the ecological impact of an activity that has long ceased in the Great Barrier Reef, particularly in the context of multiple impacts and environmental changes, it is nonetheless likely that some coral reef areas were substantially transformed by coral mining and they now exist in a highly degraded condition. Other coral reef areas experienced less intensive, yet nevertheless significant, modifications.

Coral mining in the Great Barrier Reef, 1900–1940

Before coral mining for the manufacture of agricultural and industrial lime commenced in the Great Barrier Reef, around 1900, lime burning was already an established practice; the earliest recorded instances of Europeans using shells or coral gathered from the Great Barrier Reef to produce lime date from the 1840s. Lime for the construction of the navigation beacon at Raine Island had been obtained in 1844 by burning *Tridacna* and *Hippopus* shells; and in 1847, at the time of settlement of Port Curtis (now Gladstone), an abundance of shells for lime-burning was reported in the locality (Lawrence and Cornelius, 1993, p4; Fitzgerald, 1982, p94). Lime was used to make mortar, but burnt coral was also used as a building material in its own right. In 1864, G. Bowen (1864, p116) informed the Royal Geographical Society of London that the creation of a new settlement at Port Albany, Cape York, was facilitated by the presence there of 'large beds of coral, of the best description for making lime'. Eden (1872) reported collecting coral to burn for lime from reefs near Cardwell, and in an early description of Queensland, A. J. Boyd (1882, p28) stated: 'The corals bordering our coasts also supply inexhaustible deposits of lime'. By 1900, the church at Fitzroy Island had been built by the Yarrabah Aboriginal Mission using coral taken from the fringing reef at the island (Figure 11.1). The use of coral as a building material, therefore, appears to have been an established practice and much larger structures were also constructed using burnt coral, such as the church at Darnley Island.[2] Apart from wood, coral was the most readily available building material for the construction of buildings on islands with fringing reefs, and it could be easily worked.

Coral mining took place in order to manufacture agricultural lime for the sugar cane farms on the adjacent coastal land; coral was mined from accessible coral reefs and cays and burnt as a cheap and chemically pure source of lime (Spencer and Meade, 1945, p132; King, 1965, pp104, 108; Kerr, 1995, pp92–4). Investigations by agricultural scientists from the QBSES in the early 1900s had demonstrated that the soils of the northern coastal sugar producing districts were very acidic as a result of lime deficiency. By applying lime, it was possible to correct the acidic condition of the soil, and the lime also contained a rich source of calcium, an

166 The impacts of coral mining

Figure 11.1 Church built from burnt coral at Kobbura outstation, Fitzroy Island, c.1900.
Source: Negative No. 43835, Historical Photographs Collection, John Oxley Library, Brisbane

important plant nutrient. Consequently, agricultural scientists recommended the application of lime in an attempt to increase sugar yields. Burnt lime was also used as a settling agent in the process of manufacturing raw sugar. In 1915, the QBSES reported that, in North Queensland, terrestrial sources of lime were expensive; Ernest Scriven (1915, p1175), the Director of the QBSES, stated:

> The price of lime in Northern sugar districts is still unduly high, and efforts are being made by many of the Farmers' Associations to open up various lime deposits and also to procure coral lime, coral sand, and shell deposits.

The following year, Scriven (1916, p1237) reported that interest in coral lime was high, and pulverising machines were already on the market. Farmers were advised to use coral fertilisers in combination with green manures and, by 1920, coral lime was being applied in the Mossman, Goondi, Mourilyan and South Johnstone areas at a cost of £3 per ton for coral sand and £4 per ton for burnt coral lime (Scriven, 1922, p1034).

Thus coral mining for agricultural lime commenced in 1900 and continued until at least 1940. During that period, at least twelve coral areas were mined in the Great Barrier Reef (Figure 11.2). An account of that activity was provided by a shell collector, who indicated that the mining of coral reefs occurred at the Barnard Islands around 1900. That informant stated:

> At the turn of the century last, coral mining was carried out in the Barnard Islands [...] and also at the mouth of the Mowbray River: Yule Point. Because of shifting sands and coastal erosion, at times extinct reef is exposed here along the shore. I think the sugar industry used this resource.[3]

The impacts of coral mining 167

Figure 11.2 Coral mining locations in the Great Barrier Reef, 1900–1940. Source: Author, compiled from archival files contained in PRV8340/1 Item 1, QSA; OHC 16, 2 September 2003; OHC 17, 2 September 2003; OHC 28, 19 September 2003

Another of the earliest operations took place at Snapper Island, near Cape Tribulation, where Jerry Doyle operated a lime kiln. The kiln was constructed in 1901 by the Mossman Central Mill Company (MCMC), which signed a contract with Jerry Doyle to provide burnt lime, and he produced 'ample supplies' of lime and fertiliser (Kerr, 1995, p93; QEPA, 2003c). The lime kiln was fired using timber from the nearby Daintree rainforest, which was transported to the island aboard the *Nellie*, and coral was obtained from the accessible and extensive fringing reef on the south-western side of the island; two archival sources describe the track that was cut to allow the firewood to be transported to the lime kiln.[4] The company opened a grinding plant to improve the quality of the coral lime, and Doyle's operation was still in progress in 1911 when the MCMC entered into another contract for lime with the Chillagoe Railway and Mines Company. Subsequently, in 1914, coral mining was carried out by Ishimoto, who was paid £2 per ton by the MCMC to deliver coral lime to the old wharf on the Mossman River (Kerr, 1995, pp93–4).

Another early coral mining operation took place near Innisfail, where E. Garner of Clump Point reported taking coral for agricultural lime from the foreshores of the Barnard Islands in 1900 and from Kings Reef in 1918; those activities pre-dated the introduction of the coral licensing system by the Queensland Government. Garner reported difficulty in taking much coral because 'we can only get on Kings Reef for about two hours at dead low water springs each day'.[5] However, that operation appears to have continued for many years; later, stating that he was too old to continue coral mining, Garner asked for the mining permit to be transferred to his son, Edward Henry Garner, who also operated at a coral area on Kings Reef during the 1930s.

Before 1920, other than Garner's permit, coral mining in the Great Barrier Reef appears to have been unregulated. Oral history evidence indicates that, by 1920, extensive coral mining had taken place at Snapper Island (Figure 11.3).[6] Coral mining at Snapper Island reef may have been continuous since the operation by Jerry Doyle; before the First World War, a German settler – possibly Albert Diehm – operated the lime kiln at Snapper Island and took coral from the fringing reef on the south-western side of the island. One informant, a farmer and recreational fisher, who recollected the mining operation stated: 'I can remember the railway lines across the reef at Snapper Island, on the south-west corner, where the spring is'.[7] The same informant also recalled:

> There was a German man there [...] until during the First World War, or just before it, and he was mining the coral off the big flats of coral there: it's mostly dead coral. He had a railway line across the reef. He would push out his little trolley, smash the coral off with a crowbar, put it in, wheel it up the reef – or had horses to pull it up – and take it up and burn it in a kiln that he had gouged out of the rocks there – and I think that is still there – chop the trees down on the island to burn them, and cook the coral down into a lime that he supplied to the Mossman Mill for settling their sugar.[8]

That informant stated that, at Mossman Mill, the settled mixture was removed as filter mud – or filter press – and spread on the cane fields. He reported that this practice continued until a terrestrial source of lime replaced the use of coral lime as a flocculant. He believed that the coral mining operation continued until the outbreak of the First World War, when the German settler was interned.[9]

Another oral history account, by a retired cane-cutter, indicates that coral mining was carried out at Snapper Island by Jim Tyrie.[10] That informant reported that large pieces of coral – that could be lifted by a man – were removed from the fringing reef using crowbars and were loaded into horse-drawn wagons. Those wagons were transported to the island along rail tracks that were laid across the

The impacts of coral mining 169

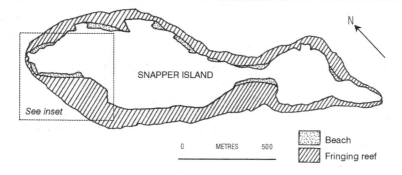

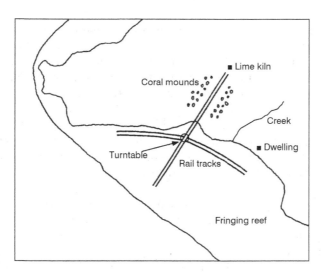

Figure 11.3 The location of the coral mining operation at Snapper Island. Source: Author, compiled from information in OHC 17, 2 September 2003; QEPA (2003c)

fringing reef, parallel to the high-water mark, and across the island to the lime kiln. A turntable was installed to transfer the wagons from one rail line to the other. The coral pieces were stored in piles beside the rail track before being burnt in the kiln and crushed; two heaps of coral and the remains of the lime kiln still survive on the island (Figure 11.3). The details of the coral mining operation at Snapper Island were described by that informant in the following terms:

> On the south-western face [of Snapper Island...] they had their lime kiln, burning the lime [...]. There was a bloke who used to live over there – this was First World War, somewhere around there, turn of the century [...] – and

he used to do the burning of the lime. They cleared a big slope of hill for firewood and it has since grown up again. [...] They had a portable tramline, like this tramline here [*indicates a nearby cane track*] [...]. The portable rail is only 20 pounds and it will take the same size wheels, so they had small trucks to cart these blocks away. There are still two big heaps of them over there that they never got around to burning. The coral was about two foot [...]: some would be a metre long. There are two big rows of them where they brought them round by boat, put them up there, and the business folded up before they could use them all.[11]

The positions of the lime kiln, the tramline and the two mounds of coral pieces are shown in Figure 11.3.

Additional details about the process of removing and burning the coral from the fringing reefs were provided by the same informant, who stated:

They had railway lines to bring the wood down [...] and they had the rail there and a turntable would come there [*indicates map*]. They dropped [the coral] into a hole and they had rail line going there and down the beach. There are big heaps of coral: a strong man would be able to pick them up and carry them. Of course, it was dead coral they got from around the fringing reef [...]; and they could go out and bust it open with crowbars, carry it back and put it in [the lime kiln]. [...] The heaps of stone are still there and, if you know where to look, you can see the big hole in the side of the hill that they used to tip this wood into and then put these stones on top so [the coral] would burn, and then they could crush it.[12]

The operation at Snapper Island pre-dated the system of coral licences introduced by the QDHM and represents a second example of unregulated coral mining in the Great Barrier Reef.

The amount of coral taken from the reef is unknown because, as an informant stated: 'He could have taken it from here for years [...]. These rails down the beach were there for a long time after the War, and they disappeared all of a sudden'.[13] However, some evidence of the scale of the operation remains in the landscape, as the informant stated:

If you went over to look at the heaps of coral, [...] you could see the heaps of stone; you could see the incinerator – the place where they burnt it – and you could see the rails, the cutting in the hill and where they had their turntable [...]. You could see all that.[14]

After Tyrie concluded mining at Snapper Island, one informant believed, he moved to the Daintree settlement and sought lime from another source. The informant suggested that a terrestrial source of lime replaced the material taken from Snapper Island reef after the lime burner ceased operating there; he stated:

'They bought lime from other sources after the bloke on Snapper Island. They started using lime from Chillagoe, which is [inland]'.[15] The evidence presented above suggests that, by that time, a considerable amount of coral had been removed from Snapper Island reef.

In contrast to the scarcity of documentary evidence for the earlier period, more extensive evidence of coral mining exists for the 1920s, by which time soil analysis had revealed the need for agricultural lime in sugar cane farming. In addition, coral mining operations had become more organised, being based on a system of coral licences. Several individuals were granted licences to remove coral for the production of agricultural lime; the survival of some of those licences makes a more substantial reconstruction of the coral mining industry possible. The existence of the licences also indicates that, by the 1920s, coral mining was taking place with the support of the Queensland Government. In 1922, Mineral Leases were issued for the removal of coral and coral sand from Green Island and from Oyster and Upolu Cays. The operations were reported to have been significant: one account claims that thousands of tons of material were removed from Upolu Cay. The licence for coral mining at Upolu Cay was re-issued in 1926, and the removal of material from those locations appears to have continued throughout the 1920s until the mid-1930s (Loch, 1991, p5; Bowen and Bowen, 2002, p291).[16] One oral history informant suggested that Upolu Cay had been mined for coral sand by the company Koppins, although the quantity of coral sand taken was unknown.[17]

Another of the pioneers of coral mining in the Great Barrier Reef was Albert Diehm of Innisfail. In 1927, Diehm was granted a Quarry Licence by the Atherton office of the Queensland Sub-Department of Forestry to remove coral from Hutchinson and Jessie Islands in the Barnard Group. During the following year, he produced lime at Maria Creek, near Innisfail, using coral from those islands. A QDHM memorandum about Diehm's operation stated that:

> The crushing works operated by [Albert] Diehm are situated on the Northern end of Hutchinson Island, North Barnard Group, above high-water mark. The plant consists of a Fordson tractor and a disintegrator. The estimated capacity is sixteen tons per day but the estimated daily output is six tons per day.[18]

At the end of 1928, Diehm applied for a Mineral Lease over half an acre of coral on the western side of Hutchinson Island and one-fifth of an acre of coral on the western side of Jessie Island in order to continue his operation.[19]

The initial success of coral mining in northern Queensland attracted the interest of capital investors in southern Australia. In 1928, an article in the *Melbourne Herald* described the industry in the following terms:

> There are splendid prospects of a profitable industry in crushing the coral of the Great Barrier Reef for fertiliser. The pioneer of the industry is Mr

Diehm, who recently installed a £500 plant on North Barnard Island, and has already supplied 200 tons of pulverised coral to Innisfail farmers. Mr Diehm stated today that one farmer had put twenty tons in his fields and the cane treated has shown an advance of two feet six inches over other cane. [...] He intends to bring regular supplies of the fertiliser to Innisfail. Recently Mr Diehm obtained additional gear from England and hopes to operate on a larger scale now that pioneering difficulties had been overcome. There were almost unlimited supplies of coral to be drawn on.[20]

By mid-1929, Diehm had extracted and crushed coral at Hutchinson Island for at least three years.

Coral was mined not only from islands and cays: it was also removed from inshore coral reefs in the northern Great Barrier Reef, which were more accessible from the mainland and more convenient to work. In 1929, a lease to mine coral at Alexandra Reef, near Port Douglas, was granted to G. Averkoff of Port Douglas who, like Diehm, intended to produce lime for sugar cane fields.[21] The location of the coral reefs was between Yule Point and the Mowbray River, and the coral lay 'approximately 5 chains below high water mark' (Figure 11.4). As the adjacent land was mangrove swamp and the removal of coral would not interfere with any other industry, the Secretary of the Queensland Marine Board, J. D. W. Dick, suggested that this application should be granted subject to a royalty of 1 d per cubic yard on all coral removed. Averkoff then constructed a lime plant and supplied coral lime to the MCMC for fifteen years, until his operation was taken over by the McDowell Brothers, who continued to deliver the lime to sugar cane farmers (Kerr, 1995, p94).[22]

Several other applications were made to mine coral during the same period. In 1929, High Island, adjacent to the Frankland Group, was the subject of a coral mining application by R. McGuigan, whose application was considered at the same time as those of Diehm. At Pialba, Henry M. Taylor stated that he had access to thousands of tons of coral and claimed the sole right to remove this material using an oil engine.[23] Companies as well as individuals made applications to mine coral. In 1929, Great Barrier Reef Fisheries Ltd (1929, p5) of Sydney proposed to manufacture 'natural fertilisers obtained from burnt coral'. In the same year, a syndicate of investors in Sydney and Melbourne applied to mine coral and limestone from seven islands in Queensland waters – including Masthead Island – in order to supply a lime works in Brisbane.[24] No evidence was found in the QSA to indicate whether or not those leases were granted.

By the late 1920s, therefore, coral mining was regarded as an industry that had the potential to generate significant profits for venture capitalists. In 1928, Edward Sanders of Cooktown applied for leases to dredge coral sand from twelve locations, comprising more than 50 acres, between Mossman and Masthead Island.[25] A syndicate formed by Sanders argued that 100,000 tons of agricultural lime could be used each year in the sugar districts – which they claimed covered 300,000 acres – and that around 10,000 tons of burnt lime were already being

The impacts of coral mining 173

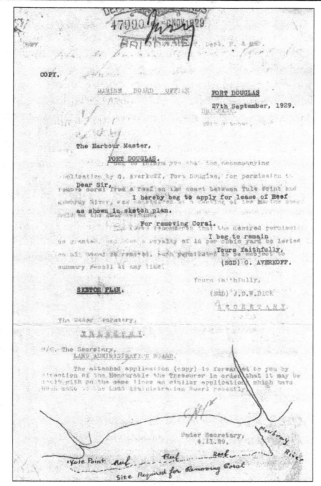

Figure 11.4 Letter and sketch map showing Averkoff's coral mining site at Alexandra Reef, 1929. Source: In-letter Ref. 47990, G. Averkoff, Port Douglas to Harbour Master, Port Douglas, 27 September 1929, PRV8340/1 Item 1, Queensland State Archives, Brisbane

used annually by sugar mills, refiners, farmers and builders.[26] The syndicate estimated the demand for agricultural lime to be 8,000 tons per year in Mackay, 8,000 tons per year in Cairns, 10,000 tons per year in Innisfail and 10,000 tons per year for burnt lime; they claimed that, at around £3 per ton, other sources of lime were too expensive for farmers. The syndicate proposed a company to work lime deposits in the Great Barrier Reef 'to supply the cane farmers with a cheap high-grade agricultural lime'; the Queensland Government Agricultural Chemist, J. C. Brunich, supported their proposal, as did Sir Matthew Nathan and the Cane Growers' Associations and Executives of Cairns, Innisfail and Mackay.[27]

174 The impacts of coral mining

Further expansion of the coral mining industry occurred during the 1930s. More extensive coral mining took place, and the industry was organised using a system of Coral Areas: reefs and cays that were individually leased and that were considered to be suitable for working. By 1930, applications by at least eight individuals and syndicates for the issue of coral licences were being considered by the Queensland Government.[28] Between 1930 and 1934, leases for five locations were granted to Edward Sanders: for Coral Areas No. 1 Cairns (Oyster Cay), No. 3 Cairns (Sudbury Cay), No. 1 Innisfail (Beaver Reef), No. 1 Mackay (Sandpiper Reef) and No. 1 Townsville (an unnamed sand cay to the north-east of Lucinda).[29] The applications for coral leases at those areas were accompanied by sketch maps, one of which is reproduced in Figure 11.5. Another Coral Area at 'Apollo Banks' (Upolu Cay) was leased to Walter Edward Tanner and Maurice Joseph Kenny of Yungaburra in 1930, whose company – Tanner and Kenny Contractors – applied to dredge for coral lime to produce fertiliser.[30] Later, in 1934, the lease for the Coral Area at Hutchinson Reef was extended and the coral leases held by Sanders, with the exception of the site at Oyster Cay, were taken up by Andrew Albert Holland of Sydney.[31]

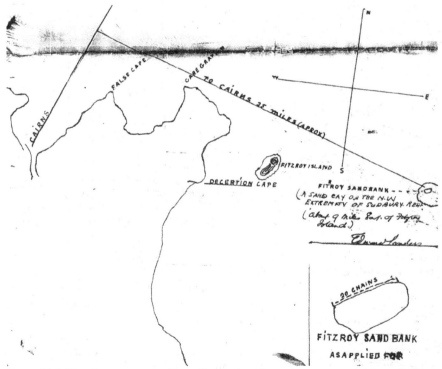

Figure 11.5 Sketch map accompanying Sanders' application to mine coral from Sudbury Cay, 1930. Sudbury Cay is referred to as Fitzroy Sandbank in this sketch map. Source: PRV8340/1 Item 1, Queensland State Archives, Brisbane

Some concerns were expressed about the advisability of permitting coral mining in the Great Barrier Reef. In addition to public complaints about the destruction of Upolu Cay, the archival sources indicate considerable differences in the opinions of Queensland Government officials towards coral mining.[32] One supporter of the industry, A. Cullen, the Chief Engineer of the QDHM, discussed the coral mining operation at Upolu Cay, stating that 250 tons of material had been removed from the cay by Tanner and Kenny during the nine-month period from 1 January–30 September 1931. Cullen argued that the public concerns about the destruction of the cay were 'sentimental' ones and that the resulting disturbance to seabirds – even if this occurred at several cays – could not be regarded seriously. Furthermore, Cullen stated: 'Assuming (by way of argument) that the cay at which material is being obtained by Messrs Tanner and Kenny was in the course of years entirely removed, it would be because a product of some value was being obtained'.[33] Cullen's view represented a utilitarian perspective towards the resources of the Great Barrier Reef; such a view – in which coral was regarded either as a source of limestone or as a means of promoting tourism – formed the basis of the coral licence system.

In contrast to the view of Cullen, in 1931, the Cairns City Council expressed its concern that the removal of coral from Green Island was threatening the popularity of the island with tourists.[34] In correspondence with the Queensland Government, the Council requested legislation to protect Green Island reef from being stripped of its coral, but this request was met with reluctance by the Queensland Treasury because coral mining was 'an industry which the Government considers it advisable to encourage'.[35] Eventually, the Queensland Government, acknowledging that there was no legal authority by which the reef at Green Island could be placed under the protection of the local Council, issued a licence for the removal of coral from Green Island reef to the Council; that licence conferred sole rights to removal of coral from Green Island reef on the Council.[36] Subsequently, in 1937, the Queensland Government did legislate to prohibit the removal of coral from the foreshores and reefs surrounding Green Island, Low Isles, Michaelmas Cay, Arlington Reef and Oyster Cay. Yet, the earlier decision to protect Green Island reef was a significant one: it created one of the earliest marine protected areas in existence (Lawrence et al., 2002, p25).[37]

Nevertheless, the removal of coral continued in other locations in the Great Barrier Reef. From 1936–1938, extensive coral mining took place in the Innisfail area. Edward Henry Garner was granted a lease over Coral Area No. 2 Innisfail (Kings Reef); Thomas Roper held a lease for the adjacent Area No. 5, and also for Areas No. 3 (Hutchinson Island) and No. 4 (Jessie Island).[38] Garner reported mining about 70 tons of coral from his site at Kings Reef during the quarter ending on 30 June 1935, and 60 tons the following quarter; he also helped to mine Roper's adjacent lease on the same reef (Figure 11.6). The licences were granted on the condition that explosives would not be used in removing coral; however, the QDHM received complaints that Garner used gelignite to blast coral from the reef before bringing the rubble ashore for burning (Jones, 1973, p317).[39]

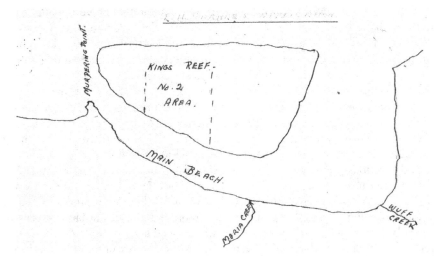

Figure 11.6 Sketch map showing Garner's application to mine coral from Kings Reef No. 2 Area, 1937. Source: PRV8340/1 Item 1, Queensland State Archives, Brisbane

In 1939, a syndicate comprising the Villalba Brothers and Martinez and Company applied for leases over Beaver and Taylor Cays, near Dunk Island, and over Coral Area No. 1 Townsville ('Sand Cay Island', to the north-east of Lucinda), in order to collect coral lime.[40] In May 1940, a lease was granted to Martinez, Chapman and Company of Innisfail to remove 1,000 cubic yards of coral from Sand Cay Island; the rights were sold for 3d per cubic yard.[41] This application indicates the willingness of investors to form syndicates to obtain coral leases and suggests that, by the end of the 1930s, coral mining in northern Queensland had become an established, profitable industry. However, the series of coral licences preserved at the QSA indicates that this lease was the last granted before the outbreak of the Second World War disrupted marine industries in the Great Barrier Reef; during the war, boats were impounded and access to the coral reefs and cays was restricted.[42]

No archival evidence was found to indicate whether or not coral mining resumed after the end of the Second World War. The reports of the Queensland Chief Inspector of Fisheries published in the *QPP* indicate that coral and shell-grit licences were issued continuously by the Queensland Government throughout the period 1930–1968, representing an increasing number of coral licences (Figure 11.7). However, those coral licences were also possibly related to the collection of coral for tourism and the aquarium trade. One oral history informant suggested that, after the Second World War, cheaper, terrestrial sources of agricultural lime were used by sugar cane farmers, including lime obtained from Chillagoe and Calcium, near Townsville.[43] In addition, in 1940, increasing attention was given to the protection of coral reefs in response to the development of tourism in the Cairns, Townsville and Whitsunday regions; the extraction of coral from 28 coral reefs in the Great Barrier Reef was prohibited.[44] Attempts

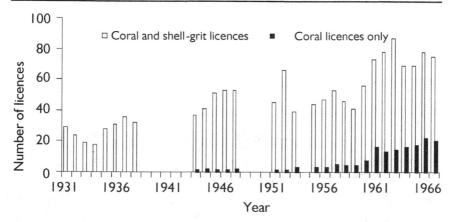

Figure 11.7 Numbers of coral and shell-grit licences issued in Queensland between 1931 and 1968. After the Second World War, a distinction was made between coral and shell-grit licences and licences issued specifically for the removal of coral. Source: Author, compiled from data provided in the Annual Reports of the QDHM, QPP, 1932–1969

were made to access additional materials relating to the use of agricultural and industrial lime from sugar industry organisations and informants in Mossman, Gordonvale, Innisfail and Brisbane; those attempts suggested that records of the activity may no longer survive. Therefore, the extent of coral mining between 1945 and 1967, when the proposal to mine coral from Ellison Reef by the Cairns District Canegrowers – an organisation representing the interests of local canegrowers – was refused by the Queensland Government, is unknown. However, coral mining probably ceased during that period.

The impacts of coral and coral sand mining in the Great Barrier Reef are difficult to assess. The earliest indication of the ecological degradation associated with industry concerned the works by Tanner and Kenny Contractors at Upolu Cay. In 1931, a complaint about their operation was published in *The Cairns Post*, which stated that Upolu Cay 'was being destroyed by a firm taking away the bank for fertiliser purposes and depriving the sea birds of a home that has been theirs for many years'.[45] Material was removed from Upolu Cay by running a tramline into the centre of the cay and quarrying coral sand to a depth of about 4 feet (Figure 11.8).[46] During the nine months from January–September 1930, Tanner and Kenny Contractors removed 250 tons of material from the cay.[47] Although their coral licence permitted the removal of coral and coral sand from the foreshore – below high water mark – Tanner and Kenny had mined coral sand from the centre of the cay and, by October 1930, the height of the cay had been reduced and almost no bird life or vegetation remained.[48]

As both Upolu and Oyster Cays had been declared sanctuaries for animal and bird life in 1926, the destruction caused at Upolu Cay provoked objections from naturalists, who were also concerned about the possibility of similar destruction at Oyster Cay. By 1933, Sanders had not yet commenced removing coral from

Figure 11.8 The jetty at Upolu Cay used for loading material mined from the cay, c.1933.
Source: PRV8340/1 Item 1, Queensland State Archives, Brisbane

his lease at that site. In spite of public protests, both leases were renewed in 1933.[49] By January 1933, Tanner and Kenny had caused further 'serious damage' to Upolu Cay – one report claimed that almost half the cay had disappeared – and continued to disregard the requirement to mine only from the foreshores of the cay.[50] In addition to the disruption caused to seabirds, the removal of material threatened the stability of the cay and increased its susceptibility to erosion during storms. Finally, in 1934, in response to complaints about the extent of destruction caused by coral mining, the coral licences for both Upolu and Oyster Cays were revoked by the QDHM (Bowen and Bowen, 2002, p291).

Other evidence of the destruction caused by coral mining exists for Kings Reef, near Innisfail, where the operation carried out by Garner also elicited complaints. Several reports claimed that the nearby bathing beach at Murdering Point had become unusable as a result of sharp pieces of coral being washed ashore after Garner's blasting operations. One of those reports stated that:

> Garner is in the habit of using explosives to loosen the coral from the reef, which when broken off he leaves in heaps. The prevailing weather and currents set in from where he is blasting towards Murdering Point beach, and the result is that sharp and light pieces of coral are washed in and are a danger to persons using the beach [...]. The pieces of coral also cut fishing nets used by the fishermen at the beach. After any boisterous weather there is always a fair amount of coral washed in to the beach, and even in fine weather a good deal of it comes in.[51]

The use of explosives by Garner was also blamed for driving fish away from the area: another concern for the fishers besides damage to their nets. On the miners

Figure 11.9 Alexandra Reef, near Port Douglas, 2003. Source: Author

themselves, the blasted corals inflicted skin burns and large 'coral sores' that resembled ulcers, which raises the possibility that those corals were alive until the explosives were used (Jones, 1973, p317).

No other documentary evidence of the destruction of coral reefs and cays as a result of coral mining was found in the archival sources. Hence, the remaining environmental impacts of coral mining can only be estimated. The nearshore coral reefs, which were the most accessible mining locations, probably suffered the most sustained and destructive impacts of coral mining. Kings and Alexandra Reefs are the reefs most likely to have been extensively degraded, since they were worked from a comparatively early date – before 1930 – and because, in comparison with the fringing reef at Snapper Island, they were more easily reached from the coast. In addition, while Snapper Island was inhabited sporadically, Kings and Alexandra reefs were accessible from population centres (including Kurrimine Beach, Innisfail, Port Douglas and Mossman) throughout the period since European settlement of the north Queensland coast. Today, both reefs appear to be almost completely degraded: the reef flat at Kings Reef is characterised by coral rubble, soft corals, mud and algae, and lacks extensive hard coral communities; the surface of Alexandra Reef, shown in Figure 11.9, comprises dead coral, with living colonies found only at the submerged edges of the reef.[52] While the dead coral found at these reefs cannot be attributed solely to coral mining, the blasting and removal of coral has probably contributed to their degradation.

Oral history evidence suggests that the impacts of coral mining were extensive at Snapper Island reef and large mounds of mined, unburnt coral still exist on the island near the remains of the lime kiln. In addition to the removal of coral from the reef flat using crowbars, the coral cover was probably diminished by trampling by people and horses, as well as by the construction of rail tracks across the surface of the reef.[53] While Snapper Island was less accessible than the nearshore reefs at Kings and Alexandra Reefs, a dwelling was built on the island, which facilitated the relatively prolonged exploitation of that reef. In common with other inshore and fringing reefs, Snapper Island reef was particularly vulnerable to human impacts. However, unlike Kings and Alexandra Reefs, in the 1990s, Snapper Island reef contained a very large cover and diversity of living corals, as one oral history informant – a marine biologist – reported. Consequently, the visible impacts of coral mining may now be more apparent in the landscape of that island than in its fringing reef.[54]

Summary

Coral mining in the Great Barrier Reef was more extensive than has previously been acknowledged. Between 1900 and 1940, coral mining developed from an isolated activity carried out by individuals into a well-organised industry, encouraged by the Queensland Government and organised using a system of licences and Coral Areas, and at least twelve areas in the Great Barrier Reef were mined for coral. Although some locations – the Barnard Islands and Kings Reef – were worked since 1900, and mining had begun at Snapper Island by 1914, more extensive operations took place during the 1920s and 1930s, which attracted syndicates of investors as well as individual coral miners. By the onset of the Second World War, coral mining had become an established and profitable industry in northern Queensland, supplying cheap agricultural lime to sugar cane farmers on the adjacent coast and industrial lime to sugar mills, and coral extraction was concentrated in the Cairns and Innisfail areas where a cheap terrestrial source of lime was not yet readily available.

This account of coral mining, however, is incomplete as a result of gaps in the archival records, the difficulty in obtaining original oral history evidence for the period before 1940 and the lack of extensive scientific monitoring of the Great Barrier Reef before around 1970. The sequence of surviving records of coral areas – which includes Coral Areas No. 1 (Cairns) and No. 3 (Cairns), but not No. 2 (Cairns), for instance – suggests that more locations were mined than are mentioned here. Furthermore, other instances of unlicensed coral mining may have taken place that are not mentioned in the documentary record, just as extensive operations took place at Snapper Island without, apparently, any documentary evidence surviving in the records of the QDHM that were consulted at the QSA. Therefore, this account gives an overview of what may have been a more extensive industry in the Great Barrier Reef. Nevertheless, the evidence presented in this chapter suggests that coral mining probably caused significant changes in at least some parts of several coral reefs. Several of the reefs that were once mined – especially Kings and Alexandra Reefs – now appear highly degraded. The ecological effects of coral mining are difficult to determine; nonetheless, all of the coral mining locations described in this chapter would probably have become more vulnerable to other environmental changes – both human and natural – and may subsequently have had a reduced capacity to recover from other environmental stresses. Some of those reefs – particularly Kings Reef – may even have crossed environmental thresholds beyond which their recovery became extremely unlikely.

Notes

1. OHC 14, 26 August 2003; OHC 16, 2 September 2003; OHC 17, 2 September 2003; OHC 28, 19 September 2003; OHC 36, 28 October 2003.
2. N. R. Strelitz, 'Trojan car trip, 1925, from Thursday Island to Pascoe River, Cape York, showing coral church (All Saints), Darnley Island', Image No. P02139, Image Library, Cairns Historical Society.
3. Anonymous, 'Recollections of the reef', Changes in the Great Barrier Reef since European Settlement, Oral History Collection, School of TESAG, JCU, September 2003.
4. In-letter Ref. AWG:LDM, William L. Rutherford, Port Douglas to District Forester, Atherton, 11 May 1967, SRS5146/1 Box 2 Item 10, NP64, Snapper Island, QSA; In-letter Ref. AWG:LDM, District Forester, Atherton to Secretary, Queensland Department of Forestry, Brisbane, 19 May 1967, SRS5146/1, NP64, Snapper Island, QSA.
5. In-letter Ref. 0663, E. Garner, Clump Point to Mr W. V. B. Forrester, Port Master, 23 January 1935, PRV8340/1 Item 1, QSA.
6. OHC 16, 2 September 2003; OHC 17, 2 September 2003; OHC 28, 19 September 2003.
7. OHC 16, 2 September 2003.
8. OHC 16, 2 September 2003.
9. OHC 16, 2 September 2003.
10. OHC 17, 2 September 2003.
11. OHC 17, 2 September 2003.
12. OHC 17, 2 September 2003; this description of the operation is similar to the account provided in OHC 16, 2 September 2003.
13. OHC 17, 2 September 2003.
14. OHC 17, 2 September 2003.
15. OHC 16, 2 September 2003; OHC 17, 2 September 2003.
16. See also 'Great Barrier Reef: Supplies of coral sand', *The Cairns Post*, 16 January 1933, p11.
17. OHC 15, 27 August 2003.
18. Memo Ref. 1159, A. E. Aitken, Harbour Master, Innisfail to Port Master, Queensland Marine Department, Brisbane, 12 March 1928, PRV8340/1 Item 1, QSA.
19. 'Application to remove coral, etc. from islands and Barrier Reef', c.1929, PRV8340/1 Item 1, QSA.
20. *Melbourne Herald*, 24 January 1928, cited in In-letter, N. G. Roskruge, Deputy Director, Navigation and Lighthouses, Queensland Marine Branch (Navigations and Lights Services), Brisbane to Dr Marks, Honorary Secretary, GBRC, Brisbane, 20 July 1929, PRV8340/1 Item 1, QSA.
21. In-letter Ref. 47990, G. Averkoff, Port Douglas to Harbour Master, Port Douglas, 27 September 1929, PRV8340/1 Item 1, QSA; In-letter Ref. 47990, E. J. Whelan, Harbour Master, Port Douglas to Chief Inspector of Fisheries, QDHM, Brisbane, 1 October 1929, PRV8340/1 Item 1, QSA.

22 Out-letter Ref. 29/9270T, J. D. W. Dick, Secretary, Queensland Marine Board Office, Brisbane to Under-Secretary, Treasury, Brisbane, 29 October 1929, PRV8340/1 Item 1, QSA
23 'Application to remove coral'; Out-letter Ref. 28/7419, Queensland Marine Department, Brisbane to Chief Engineer, Queensland Harbours and Rivers Department, Brisbane, 24 April 1928, PRV8340/1 Item 1, QSA.
24 J. E. Lane, Brisbane to Secretary, Queensland Provisional Forestry Board, 18 October 1929, PRV8340/1 Item 1, QSA.
25 In-letter, E. Sanders to Minister for Mines, Brisbane, 22 March 1928, PRV8340/1 Item 1, QSA; In-letter, E. Sanders to Minister for Mines, Brisbane, 4 April 1929, PRV8340/1 Item 1, QSA.
26 In-letter, E. Sanders to Minister for Mines, 15 August 1928, PRV8340/1 Item 1, QSA.
27 In-letter, Messrs E. Sanders and others' syndicate (E. Sanders, G. H. Pritchard, Jas. G. Campbell and T. L. Jones) to A. J. Jones, Minister for Mines, Brisbane, 7 December 1928, PRV8340/1 Item 1, QSA.
28 'Regulations governing the taking of coral or shell-grit – Section 18 of *The Fish and Oyster Act of 1914*', PRV8340/1 Item 1, QSA.
29 These details were compiled from many files contained in PRV8340/1 Item 1, QSA.
30 In-letter Ref. 29/2120, Tanner and Kenny Contractors to Secretary, Queensland Department of Mines, Brisbane, 5 August 1929, PRV8340/1 Item 1, QSA; In-letter Ref. 30/6493, Forbes, for Under-Secretary, Brisbane to Chief Inspector of Fisheries, QDHM, Brisbane, 13 November 1930, PRV8340/1 Item 1, QSA.
31 In-letter Ref. 6699, A. A. Holland, Sydney to Chief Inspector of Fisheries, QDHM, Brisbane, 30 October 1934, PRV8340/1 Item 1, QSA; Out-letter, Chief Inspector of Fisheries, QDHM, Brisbane to Harbour Master, Mackay, 5 October 1934, PRV8340/1 Item 1, QSA; In-letter Ref. 33/5239, A. A. Holland to Chief Inspector of Fisheries, QDHM, Brisbane, 1933, PRV8340/1 Item 1, QSA.
32 F. H. Dean, 'Correspondence: destroying Upola Bank on Barrier Reef', *The Cairns Post*, 27 October 1931, p. 11; In-letter, F. H. Dean, Kuranda to Mr Atherton, Minister, Queensland Department of Mines, Brisbane, 20 November 1931, PRV8340/1 Item 1, QSA.
33 Out-letter Ref. 31/9363, A. Cullen, Chief Engineer, QDHM, Brisbane to Under-Secretary, Treasury, Brisbane, 19 November 1931, PRV8340/1 Item 1, QSA.
34 In-letter, Town Clerk, Cairns to Queensland Marine Board Office, Brisbane, 21 August 1931, SRS146/1 Item 2, QSA.
35 In-letter Ref. L.A.C.T. Gen., Under-Secretary, Treasury, Brisbane to Town Clerk, Cairns, 22 January 1931, SRS146/1 Item 2, QSA.
36 In-letter, Under-Secretary, Treasury, Brisbane to Town Clerk, Cairns, 20 April 1932, SRS146/1 Item 2, QSA.
37 SRS146/1 Item 2, QSA.
38 In-letter Ref. 37/4439, Thos. G. Hope, Acting Under-Secretary, Treasury to Official Secretary to the Lieutenant Governor, Brisbane, 8 July 1937, SRS31/1 Box 13, QSA.
39 In-letter Ref. 0991, E. H. Garner, Clump Point to Inspector of Fisheries, Innisfail, 29 January 1936, SRS31/1 Box 13/1, QSA
40 Out-letter, Secretary, Provisional Administration Board, QDHM to Secretary, Land Admin. Board, Queensland Department of Public Lands, Brisbane, 31 July 1939, SRS31/1 Box 13/1, QSA.
41 In-letter Ref. 225/47, Secretary, Queensland Forest Service, Queensland Forestry Sub-Department, Brisbane to Secretary, QDHM, Brisbane, 26 July 1940, SRS31/1 Box 13/1, QSA.
42 OHC 42, 13 November 2003.

43 OHC 10, 10 March 2003.
44 Letter Ref. 4868, Secretary, QDHM, Brisbane to Secretary, Queensland Sub-Department of Forestry, Brisbane, 29 October 1940, PRV8340/1 Item 1, QSA.
45 F. H. Dean, 'Correspondence: destroying Upola Bank on Barrier Reef', *The Cairns Post*, 27 October 1931, p. 11; In-letter, F. H. Dean, Kuranda to Mr Atherton, Minister, Queensland Department of Mines, Brisbane, 20 November 1931, PRV8340/1 Item 1, QSA.
46 In-letter Ref. 33/3117, C. J. Hamilton, Land Agent and Deputy Land Commissioner, Cairns to Secretary, Land Administration Board, Brisbane, 19 January 1933, PRV8340/1 Item 1, QSA.
47 Out-letter Ref. 31/9363, Chief Engineer, QDHM, Brisbane to Under-Secretary, Treasury, Brisbane, 19 November 1931, PRV8340/1 Item 1, QSA.
48 In-letter Ref. 0316, J. Brewster, Harbour Master, Cairns to Port Master, Brisbane, 16 January 1932, PRV8340/1 Item 1, QSA
49 In-letter Ref. 0316, J. Brewster, Harbour Master, Cairns to Land Agent, District Land Office, Cairns, 30 August 1933, PRV8340/1 Item 1, QSA.
50 In-letter, C. J. Hamilton, Land Agent, District Land Office, Cairns to Secretary, NQNC, Cairns, 7 September 1933, PRV8340/1 Item 1, QSA; In-letter, C. J. Hamilton, Land Agent, District Land Office, Cairns to Mr H. F. Todd, Assistant Secretary, GBRC, 28 September 1933, PRV8340/1 Item 1, QSA; In-letter, H. J. Freeman, Instructor in Fruit Culture to Under-Secretary, QDAS, Brisbane, 12 September 1933, PRV8340/1 Item 1, QSA.
51 In-letter, Constable F. R. Donovan, Silkwood to Police Magistrate, Innisfail, 13 July 1935, SRS31/1 Box 13/1, QSA; In-letter Ref. 6174, Inspector of Fisheries, Innisfail to Chief Inspector of Fisheries, QDHM, Brisbane, 10 August 1936, SRS31/1 Box 13, QSA.
52 OHC 35, 20 October 2003.
53 OHC 16, 2 September 2003; OHC 17, 2 September 2003.
54 OHC 20, 9 September 2003.

Chapter 12

Other impacts on coral reefs

Introduction

This chapter presents evidence of some other physical impacts on coral reefs of the Great Barrier Reef that have occurred since European settlement: the clearing of access channels and tracks in coral reefs, the effects of military target practice, the impacts of reef-walking and the development of tourism infrastructure, which are discussed below in turn. All of those activities commenced – and some had ceased – before the formation of the GBRMP in 1975. Some of the activities mentioned in this chapter – such as military operations and the clearing of access channels and tracks – had highly destructive yet localised impacts on coral reefs. In contrast, other activities, such as reef-walking, were more widespread and prolonged before protective legislation was introduced to restrict them. Therefore, the activities presented in this chapter caused a wide range of impacts that varied in their geographical distribution and intensity. The evidence presented in this chapter indicates that the reefs at Heron, Lady Musgrave, Fairfax and Green Islands, North Reef and Upolu Cay, in particular, have been significantly degraded by some of these activities.

The creation of access channels and tracks in coral reefs

Some instances of damage to corals in the Great Barrier Reef as a result of the creation of access channels and tracks have been recorded, including descriptions of the large boat channel and harbour formed at Heron Island, the boat channel created at Lady Musgrave Island and the access tracks cleared to allow the servicing of lighthouses. The access channel created at Heron Island is the largest of those channels; it is also the example for which most documentary and oral history evidence exists. An early attempt to improve access to Heron Island for boats took place in the early 1960s, when explosives were used to breach the outer rim of the reef on the south-western side of the western tip of the cay, close to the wreck of the *Sydney*.[1] Then, in 1965, Queensland Airlines proposed a Sandringham flying-boat service to Heron Island and applied for permission to build a sea-plane landing strip in the Heron Island lagoon. The creation of the landing strip required the removal of around thirty-five coral 'bommies' (micro-atolls) from the proposed landing area.[2] The Director of the QDHM, A. J. Peel, wrote to the Queensland Treasury stating that there would be no objection to the removal of the bommies in the lagoon 'provided that any necessary blasting is kept to a minimum and small charges are used'.[3]

Subsequently, between October 1966 and October 1967, the channel and harbour at Heron Island were dredged to allow easier access for boats across the reef to the cay. The dredge spoil was used to create a bank around the boat channel in an attempt to prevent sediments washing into the depression; spoil was also deposited as a spit on the south-western side of the island.[4] The channel altered the appearance of the Heron Island reef, which was photographed by the marine scientist, Isobel Bennett, before and after the creation of the channel; two of her photographs are reproduced in Figure 12.1. Following the creation of the channel, concern was expressed about rapid erosion as a result of changing sediment flows over the Heron Island reef. Erosion was reported at the western end of the island from 1960 to 1966: the period since the initial breach in the outer rim of the reef was made.

However, the full impacts of the channel were not immediately discernible, as a report about the impacts of the creation of the boat channel, written by Patricia Mather, Honorary Secretary of the GBRC, in 1971, stated:

> The effects of the most recent activity – the cutting of a channel through the reef crest at the south-west end of the cay and the excavation of a harbour with half-tide walls cannot yet be evaluated. But build-up of sand along the southern side of the cay – where it was previously being lost – and loss of sand around the north-west and western parts appears to be taking place rapidly as a result of the change in flow characteristics past the island and over the reef, caused by the presence of this deep channel through the reef.[5]

Inevitably, however, a significant area of the coral reef was affected by the construction of the channel. Another report, in 1970, claimed that 'virtually no recolonisation' of corals had occurred since the creation of the channel (Royal Commission into Exploratory and Production Drilling for Petroleum in the Area of the Great Barrier Reef, 1974, p724). In addition to those reports, many oral history informants observed changes at Heron Island reef associated with the dredging of the boat channel, especially changes in sedimentation in the channel and in the surrounding portions of the reef flat.[6] One stated that near the channel, adjacent to the island, 'the entire top of that reef dropped probably in the order of four centimetres [...] because of the speed of draining of the lagoon that used to occur at that end'.[7]

Other channels and tracks were created in reefs besides the channel at Heron Island. Another boat channel was cut through the reef at Lady Musgrave Island; at that reef, a report by Steers (1938, p56) during a geographical expedition to the Great Barrier Reef stated:

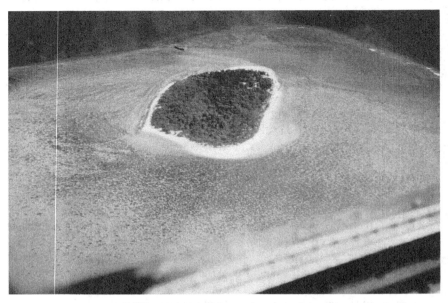

Figure 12.1 (a) The Heron Island reef before the construction of the boat channel, 1948 (top); (b) the same reef after the construction of the boat channel, August 1971 (bottom). Source: Photographs taken by Isobel Bennett, used with permission

Other impacts on coral reefs   187

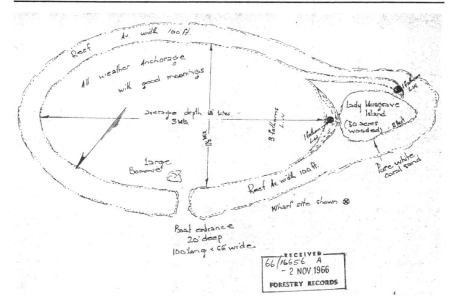

Figure 12.2 The boat channel created at Lady Musgrave Island reef, 1966. Source: Ref. 66/16656A, 2 November 1966, SRS5416/1 Box 9 Item 57, NP224, Bunker – Lady Musgrave Island, Queensland State Archives, Brisbane

There is a narrow passage through the reef which is said to have been made by Japanese fishermen. I have no definite information on this matter, the passage is certainly narrow, and as far as appearances are concerned could have been formed in this way. As it is the only clear gap through the reef, and contains reasonably deep water, it is not easy to explain it on purely natural grounds.

A survey undertaken in 1966 recorded the position and dimensions of the boat channel at Lady Musgrave Island reef: the channel was located to the northwest of the cay, and was 100 feet long, 66 feet wide and 20 feet deep (Figure 12.2).

Another access track was created at North Reef, in around 1960, to allow the lighthouse supply vehicle – an amphibious DUKW vehicle – to transport stores from the *Cape Moreton* supply vessel to the lighthouse. An entry in the *Sailing Directions* used by the Captain of the *Cape Moreton*, made on 13 May 1960, stated: 'Narrow gap in live coral to be blasted to width suitable for [low water] DUKW landing' (Chesterman, 1973, no pagination). The track was created at the edge of reef flat, on the north-western side of the island (Figure 12.3). Many other reefs were traversed by the amphibious vehicles used by the lighthouse supply service, and the *Sailing Directions* describe the difficulty in negotiating some reefs in the vehicles as a result of isolated coral outcrops and coverings of live soft corals, which presumably were damaged in the process (Chesterman, 1973).

188   Other impacts on coral reefs

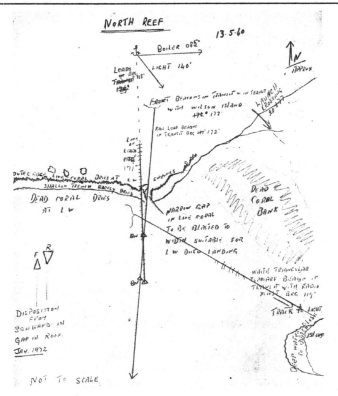

Figure 12.3 The access track for the lighthouse supply vessel at North Reef, 1960. Source: Chesterton (1973, no pagination), obtained from the Queensland Museum, Brisbane

Military impacts in the Great Barrier Reef, 1940–1960s

Some documentary and oral evidence suggests that some coral reefs were damaged by military activities, especially the reef areas that were used for bombing practice. The impact of military activities was greatest around the time of the Second World War, when mine-laying took place in the Great Barrier Reef, and in the two decades afterwards, when several islands and reefs were used for military target practice. In 1940, the threat of Japanese invasion from the Coral Sea prompted the Australian Navy to lay mines in each major shipping passage through the Great Barrier Reef; the No. 11 Catalina squadron, based at Cairns, was responsible for long-range mine-laying operations in the Great Barrier Reef (Bartlett, 1940, p7; Baglin and Mullins, 1969, p32). The impact of the mines used in the Second World War lasted beyond the duration of that conflict; Lurie (1966, pp79, 81) described the discovery of an unexploded bomb during the 1960s at Michaelmas Cay, where a controlled detonation of the bomb was carried out by the Australian Navy. Several oral history informants recalled the mine-laying and the occasional explosions of mines that drifted onto coral reefs; one informant believed that an explosion of a Second World War mine, located by the Australian Navy, occurred at Mackay Reef, and another recalled the sinking of the *Warrnambool* in Princess Charlotte Bay while attempting to retrieve mines after 1945.[8]

One oral history informant described the explosion of a mine at Green Island, in around 1946, in the following terms:

> The remnants of World War Two [...] were visible everywhere. Mines, the big brown balls with all the spikes poking out of them; some were sunk on the edge of the reefs, some were on top of the reefs, some were washed offshore on sand cays and even one, in about 1946, drifted up one night on the south-eastern side of Green Island on a high tide. It hit the rocks and exploded. [...] There was a building there that they called the kiosk: it blew the front off this.[9]

The same informant reported that the mines sometimes escaped from the chains that held them in place in the shipping lanes; he argued that the mines 'would have damaged the reef [...] pretty severely, because they were big bombs'. However, in addition to mine-laying, the Catalina aircraft also took part in target practice, during the Second World War, at the reef at Upolu Cay; the informant reported that, after the bombing of the reef, 'there was shrapnel all around the place'.

Few details of military activities were found in documentary sources; for example, although Cid Harbour, at Hayman Island, was used as a submarine base during the Second World War, no descriptions of the impacts of the base were found. However, some evidence of the effects of bombing practice in the Great Barrier Reef after the Second World War exists, because some observers visited the target sites afterwards and reported on the damage inflicted there. In 1952, the Chief Inspector of Fisheries, E. J. Coulter (1952, p1011) stated that:

> A trip to Lady Musgrave Island to collect specimens of fish and coral revealed that this reef is now practically a marine desert, which, in all possibility, is attributable to the fact that the area was used as a practice bombing target during the war.

In addition to the extensive damage at Lady Musgrave reef, further destruction took place at Fairfax Island, which was also used as a bombing range. For Fairfax Island, details about the destruction of corals were not found, but the damage to corals – as at Lady Musgrave Island – was probably severe, as Hopley (1982, p341) has acknowledged.

Military target practice, involving bombing and shelling, occurred at Fairfax Island (a double island) during the period 1943–1965, with the island itself used as a target. Consequently, there were changes in the morphology, vegetation and seabird populations of the island. One report by a National Parks Ranger, describing the condition of the island before the Second World War, stated that:

> One section of Fairfax Island was thickly timbered with *Pisonia umbellifera*, *Pandanus pendunculatus*, and *Casuarina* (Oak); [...] bird life is very plentiful on this island and during the time of his visit the brown gannet were nesting there in thousands; and [...] the island is also the nesting place of the mutton bird during the nesting season.[10]

Yet when the Queensland Government Ichthyologist, Mr T. C. Marshall, visited Fairfax Island shortly after 1945, he commented that the gannet rookery there 'was not one tenth its pre-war size when you could hardly move among the thousands of nests without stepping upon one of them', and he attributed the decline of the bird population to naval bombing practices during the war.[11]

In October 1953, another report, by C. Roff, described the effects of bombing practice at Fairfax Island. Roff stated that:

> Large numbers of the brown gannet, *Sula leucogaster*, are breeding on the island [...]. Birds continue to sit on nests whilst aircraft roar overhead and rockets explode. [...] The gannets still extensively use the island although it has been used as a target area since 1943, and in some instances, apparently during the war years, was actually bombed and shelled. (This is evidenced, in the aerial photograph attached, by the old craters on the right end of the larger of the two islands).[12]

Although Roff did not observe evidence of disturbance to the sea-bird populations, his report refers to changes to the morphology of the islands: the craters formed as a result of the bombing and shelling. The disturbance caused to the islands by bombing and shelling – and evidence of the repeated breaching of Fairfax Reef – has been acknowledged by Hopley (1982, pp341–2).

Another account of the impacts of bombing at Fairfax Island contains additional evidence of the formation of large bomb craters and the destruction of vegetation; that report of April 1954, by Mr D. Jolly, contained the following description:

> As the result of the bombing and shelling of the National Park by the Navy, there are some large shell craters at the eastern portion of the island in which an elephant could be buried. Fortunately this area is treeless. On the western portion of the west island are some bomb craters near and among the trees. After the attack on the island by the Navy the trees were almost stripped of leaves.[13]

Further bombing of Fairfax Island occurred in August and September 1963, and notification of further bombing and shelling exercises was issued to mariners in November 1965.[14] By that date, Fairfax Island had been bombed and shelled for target practice for more than two decades, and considerable damage to the landscape of the island must have been sustained.

Impacts of reef-walking

Some coral reefs have been heavily used for reef-walking by tourists (Figure 9.4). Reef-walking is not only a recent phenomenon, as Denton (1889, p171), stated:

> We spent hours wandering over [the coral reefs] at low tide. They extended partly round the island, and were a constant pleasure and delight to us. It seemed a shame to walk over the reefs, breaking at every step lovely corals, which would be the pride of our museums. Some of the branching corals, radiating from a centre, and as large over as a round table, were very graceful. [...] In places, the reef was covered [...] with soft corals – or 'sea flesh,' as it is called – resembling thick, wet leather, and very smooth and slippery to walk upon.

Another account referred to the 'deep indigo of *Heliopora* coral as the foot snaps it' (Bedford, 1928, no pagination). Many visitors to the Great Barrier Reef went reef-walking, and in 1932 the QGTB (1932, no pagination) issued the following advice to reef-walkers: 'Old boots should be carried for use when walking in the lagoon, also boots to protect the feet and ankles from coral scratches'.

Ironically, while damage was inflicted on the corals by reef-walking, the individuals who visited the reefs provided some of the only reef descriptions for the period before underwater observations using snorkels and SCUBA equipment were possible. For example, an account of dead and living corals at Cape Tribulation reefs, written by Joske (1930, p180), was based on his experiences of reef-walking. Hence, a quandary existed in relation to the value of reef-walking: the activity damaged corals, yet also allowed some individuals to gain knowledge about the nature and diversity of coral reefs, and reef-walking became a popular activity that appealed to naturalists, scientists, 'beachcombers', and coral and shell collectors. Geographically, however, the impacts of reef-walking were concentrated at the major tourist resorts, especially Green and Heron Islands; from those cays, visitors could easily access large expanses of coral reef. As early as 1938, one report stated: 'Parties of sightseers are frequently guided over the Green Island reef at low water', and a similar intensity of use of the reef flat, by reef-walkers, was observed at Heron Island (Ratcliffe, 1938, p139; see also Gunn, 1966, p109).[15]

After the formation of the GBRMP in 1975, the threat presented by reef-walkers to the most popular coral reefs was acknowledged; Woodland and Hooper (1977), for instance, found that rapid alterations to coral reefs could occur as a result of reef-walking at popular locations such as Heron Island reef. A report about the degradation of corals at Green Island, published in 1978, stated: 'Under certain conditions reef walking can be very destructive. The greatest damage occurs in very fragile habitats but can also be significant where the concentration of reef walkers is very high' (Goeden, 1978, p12). A submission by the Queensland Conservation Council, also in 1978, argued the need to 'disperse areas of reef walking' in order to minimise the damage occurring at the major tourist centres, especially Green and Heron Islands.[16] As the impacts of reef-walkers were concentrated spatially and temporally, one report about Green Island stated that:

> Restrictions should be placed on people walking on the reef at low tide. Numbers are large when low tides coincide with the times of day visits during peak periods, and damage caused simply by walking on the reef must be significant.[17]

Although it is not possible to reconstruct precisely the impacts of reef-walking, which was an informal and widespread activity, the evidence presented briefly above suggests that reef-walking, like many other activities in the Great Barrier Reef, pre-dates the establishment of the GBRMP by many decades, and has prompted significant concerns about its contribution to the degradation of some reefs.

Impacts of tourist infrastructure development on coral reefs

Dramatic changes have occurred in numerous reefs and islands of the Great Barrier Reef as a result of the development of tourist infrastructure, including jetties and tourist resorts. The development of that infrastructure reflects the expansion of tourism in the Great Barrier Reef, especially since around 1930. Some of the major tourist resort islands, including Green, Magnetic and Hayman Islands, have experienced the most obvious transformations, especially in the views of some oral history informants.[18] Extensive documentary evidence also describes the construction of jetties at Green, Magnetic and Molle Islands. The earliest jetty at Magnetic Island was constructed at Picnic Bay in 1902 by Robert Hayles, in order to facilitate the landing of tourists at the island from his launch, the M.V. *Phoenix*; that jetty was destroyed by Tropical Cyclone Leonta in March 1903 and a replacement was constructed. Subsequently, jetties were also built at Nelly Bay and at Arcadia; both of those were destroyed in cyclones and rebuilt in about 1927. Damage to the later jetties occurred in another cyclone, in 1940, and replacements were built (O'Donoghue, 2001).

At Green Island, the first jetty was built by the Cairns Harbour Board in 1906, although regular commercial launches to Green Island did not commence until 1928. Shortly afterwards, in 1931, a second jetty was built at the island by the Cairns Town Council to facilitate the construction of 'other improvements' at the Green Island tourist resort, including a kiosk and an accommodation block, which were completed by 1936. In 1938, Charles Hayles, of Hayles Magnetic

Figure 12.4 A jetty constructed at Green Island, c. 1956. Source: Image No. P55090, Image Library, Cairns Historical Society, Cairns

Pty Ltd, was granted a lease to develop Green Island as a tourist resort; that lease involved the provision of regular public ferry services to the island. In 1946, the jetty at Green Island was destroyed in a cyclone and was reconstructed by the Cairns Harbour Board; the new jetty, illustrated in Figure 12.4, was replaced by another in 1960–1961, also built by the Cairns Harbour Board (Baxter, 1990; QEPA, 2003a).[19] In addition to the jetties at Green and Magnetic Islands, another jetty had been built at Molle Island by 1969, which extended as far as the outer edge of the reef.[20] The construction of jetties at those islands was significant because the structures facilitated access to, and further development of, the islands and their reefs.

Tourist resorts were constructed on many islands of the Great Barrier Reef. Before around 1900, the relative inaccessibility of many islands had precluded their use by large numbers of visitors. However, subsequently the islands became increasingly popular as tourism destinations. Davitt (1898, p282) described their attraction in the following account:

> There is no other coastal scenery in the world to equal this in changing vistas of loveliness and grandeur. You move along in endless windings in and around the islands of coral, with their silvery sands and grassy slopes, and wooded vesture of varied foliage. I journeyed by night in one trip, and by day in another, through this enchanted world of coral islands, and had a double enjoyment of its scenic beauties.

Davitt (1898, p282) also described the Great Barrier Reef islands as 'Nature's necklace of coral islands', although even in his account the islands were viewed from a passing vessel rather than regarded as destinations in their own right.

The earliest island tourist resorts included Double, Green and Magnetic Islands, which were more accessible from the coastal centres at Cairns and Townsville than the offshore islands. In particular, the resorts at Green and Magnetic Islands, which were serviced by the Hayles Magnetic Pty Ltd ferries, were frequently visited. By 1925, the Green Island resort had become so popular with visitors that W. W. M. McCulloch, the Superintendent of the Yarrabah Aboriginal Mission, expressed concern that the island was being degraded. In a report to the Queensland Chief Protector of Aboriginals, he stated:

> Green Island at present owing to excursionists is like a sewerage farm and I have been told that several influential Cairns folk, including the Mayor […] and some of the Councillors, who have always spent their holidays there, want to go to Turtle Bay this year instead.[21]

Besides Green Island, the early development of tourist resort islands also took place in the Whitsunday Group. In 1935, A. Busuttin, the Lessee of Brampton Island, stated that a small village resort, with accommodation for 50 or 60 people and other facilities, had been constructed on that island since about 1930.[22]

More extensive construction of tourist facilities occurred in 1938 and 1939 at Lady Musgrave Island, in the Capricorn-Bunker Group, and at various Whitsunday Islands. In 1938, H. F. Baker reported that the construction of a tourist resort, including six cottages and other facilities, was about to commence at Lady Musgrave Island.[23] In the same year, Caldwell (1938) acknowledged the popularity of the fishing grounds around the Whitsunday Islands, several of which were being developed as tourist resorts. By 1939, however, E. O. Marks, the Honorary Secretary of the GBRC, acknowledged that some islands of the Great Barrier Reef – especially in the vicinity of the tourist resorts – had been degraded by the thoughtless behaviour of tourists.[24] In the same year, C. J. Trist, the Secretary of the Queensland Sub-Department of Forestry, wrote:

> Despite the protective measures that are already in existence to preserve the natural beauty of the islands of the Barrier Reef, complaints are still being made that vandalism occurs. Thoughtlessness rather than vandalism can better describe the desire of this temporary [tourist] population to souvenir and interfere with the natural beauty of these islands.[25]

By 1940, the impacts of tourism in the Great Barrier Reef were concentrated in the vicinity of ten major resort islands, from Heron to Green Island, as shown in Figure 12.5. Lady Musgrave and Heron Islands were accessed from Bundaberg and Gladstone respectively; the Whitsunday Islands were reached from Mackay and Proserpine; Magnetic Island was reached from Townsville; Dunk Island services departed from Tully; and Green Island was accessed from Cairns (QGTB, 1940, p3).

Other impacts on coral reefs 195

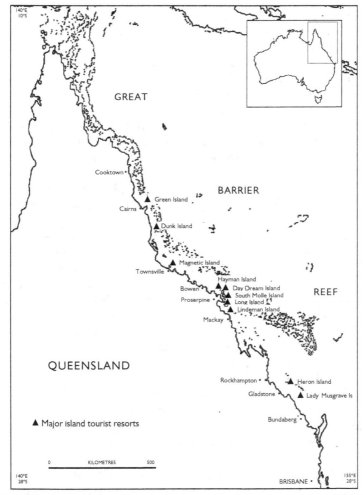

Figure 12.5 The ten major island tourist resorts of the Great Barrier Reef, 1940. Source: Author, compiled from information provided in QGTB (1940, p3)

Another account of environmental degradation at a tourist resort island was produced in 1940, when an officer of the QGTB visited Lady Musgrave Island; that officer reported that tourists had caused damage to corals, turtles and other wildlife at that island. In addition, the impacts of tourists spread beyond the resort islands to include nearby islands; G. Gentry acknowledged the degradation of Hoskyn and Fairfax Islands as a result of frequent visits by tourists from the Lady Musgrave and Heron Island resorts.[26] Nevertheless, tourism continued to develop in the Great Barrier Reef. In particular, the Green Island resort expanded considerably, due to the popularity of glass-bottom boat tours, which commenced in 1937 and which were probably the first ever example of that activity. By the early 1940s, Green Island had been declared a National Park, the visitor facilities and camping areas at the cay had been developed and the Coral Cay Hotel had been constructed.[27]

By 1948, the tourist resort at Heron Island had also expanded and comprised about ten cabins in addition to other facilities; an inspection of the resort during that year by the Director of the QGTB revealed that:

> The island has enjoyed a wide popularity with tourists over a period of about fifteen years, because of its excellent geographical situation, its excellent coral, fishing and opportunities to see Barrier Reef life.[28]

The growth of tourism at Heron Island was facilitated by a twice-weekly air service from the Brisbane River to the Heron Island resort, operated by Barrier Reef Airways using a Catalina flying-boat, which commenced in about 1947; before that year, tourists travelled by rail from Brisbane to Gladstone before making a six- or seven-hour crossing by boat from the mainland to the cay. In addition to the service to Heron Island, Barrier Reef Airways also provided weekly flights to Lindeman and Day Dream Islands, both in the Whitsunday Group, and in the first two months of its operation, Barrier Reef Airways carried around 1,000 passengers to the Barrier Reef (Bartlett, 1947, pp6–9).

Further expansion of tourist facilities on Great Barrier Reef islands occurred in the 1950s. By 1950, Dunk Island had become a popular resort island, as National Parks Ranger McKeown acknowledged:

> Dunk Island is one of the major, and most popular, tourist resorts of North Queensland, where fish and oysters are largely featured. The fact that they can wander over a reef, and pick up a feed of oysters as they go, has a strong appeal to many southern, and overseas tourists.[29]

One oral history informant, a charter boat operator, stated that the resort at Dunk Island was owned by Australian Airlines. At the resorts in the island National Parks, another oral history informant – an official of the Queensland Forestry Department – explained, the facilities were developed and then the resorts were excised from the National Parks; the resort owners were also issued with Special Leases for the remaining parts of the islands in order to prevent other companies from establishing competing resorts on the same islands. Access to the resort at Dunk Island was facilitated by a daily ferry service; the ferry operator also offered glass-bottom boat tours to Bedarra Island, at which another resort was located.[30]

During the 1950s, significant developments took place at Green Island. In 1954, the Coral Gardens underwater observatory was constructed at the end of the jetty, following a successful application by Vincent Vlasoff and Lloyd Grigg, who were issued with Special Lease SL25283 over a popular area of the coral reef. National Parks Ranger McKeown described the location of the observatory in the following terms:

The most attractive coral garden at Green Island is located adjacent to the jetty, on its northern side. Visitors to the island are conveyed over this coral in a large flat bottom boat, into which glass windows have been fitted, allowing a good view of coral, and other marine life on the ocean bed. A small, but very beautiful patch of coral is located close to the jetty, on the southern side, and is easily viewed from the jetty deck. This coral patch never fails to attract the attention of all visitors.[31]

The observatory, a 70-ton cylindrical steel chamber with glass windows, was floated from Cairns harbour to the end of the Green Island jetty, where it was sunk and anchored. The coral gardens surrounding the observatory were supplemented with coral that had been transplanted from the surrounding reefs (see Chapter 9). In 1955, the observatory commenced operations: from 1957–1958, 28,000 people visited Green Island. In 1960, further expansion of accommodation and other facilities commenced, and 47,646 visitors to the island were recorded.[32]

Considerable development of infrastructure also took place at Hayman Island during the 1950s. The island was leased by Barrier Reef Islands Pty Ltd, and the main attraction for visitors was the large fringing reef on the southern side of the island, of which more than 1,000 acres was exposed at low tide. A hotel was constructed on the foreshore of a sheltered bay on the south side of the island.[33] The owners of the Hayman Island Resort subsequently applied for a large area of the island to be excised from the National Park and leased to them; despite a recommendation to the contrary by the Queensland Conservator of Forests, the Hayman Island National Park – which comprised 1,758 acres – was cancelled during the year 1959–1960, as oral history evidence indicates, and a resort was constructed on the leased land (Figure 12.6).[34]

The expansion of island tourist resorts led to further concerns about various forms of environmental degradation. In 1966, at Lady Musgrave Island, Curtis reported that a large proportion of the coral inspected on the reef was dead and, in 1969, V. B. Sullivan, the Queensland Minister for Lands, also wrote that the coral at Lady Musgrave Island was poor and greatly inferior to the less-visited Heron Island reef. Oral history evidence indicates that concerns were raised about the environmental impacts of tourism at the Lindeman Island Resort during the late 1960s, and also at Green Island. At the latter island – which by 1975 was visited by over 100,000 people annually – the effects of sewerage outfall from the resort and amenities were reportedly causing the proliferation of seagrass beds in areas where coral reef flats had previously grown.[35] By 1978, when an estimated 150,000 people visited Green Island, G. Goeden of the Queensland Fisheries Service acknowledged that the marine resources of the island had been depleted.[36] Additional evidence of changes in islands due to tourist developments exist for the later period; those changes included the construction of the tourist resort at Lizard Island, which had been completed by 1995, and the extensive modification of the spit at Dunk Island, by 2000, among other schemes (QPWS, 2000, p10).[37] Overall, by the time of the formation of the GBRMP,

Figure 12.6 Construction of the Hayman Island Resort, c. 1962. Source: Photograph Ref. 3273, SRS189/1 Box 17 Item 73, Queensland Industry, Services, Views, People and Events; Photographic Proofs and Negatives; Islands – Barrier Reef, Queensland State Archives, Brisbane

the construction of tourist infrastructure in the Great Barrier Reef, as narrated above, had resulted in dramatic changes to the landscapes of some islands, and probably also to their reefs, particularly at Green, Hayman, Magnetic, Dunk and Lady Musgrave Islands.

Summary

This chapter has presented evidence of various other physical impacts on the reefs of the Great Barrier Reef (besides those described in earlier chapters). The clearing of access channels and tracks in coral reefs caused significant changes in the structure of the reefs at Heron and Lady Musgrave Islands, and at North Reef. Military impacts – including the bombing and shelling that occurred for the purpose of target practice – caused profound destruction at several locations, particularly the reefs and islands of Lady Musgrave and Fairfax Islands and Upolu Cay. Reef-walking was an activity with smaller individual impacts, but cumulatively was a prolonged and widespread activity in the Great Barrier Reef, and particular concerns were expressed about the damage it caused to reefs at Green and Heron Islands. The development of tourist infrastructure on some of the islands of the Great Barrier Reef also led to localised degradation, and again particular concerns focused on the degradation of Green, Heron and Lady Musgrave Islands, as well as some other popular destinations such as Hayman Island. The impacts described in this chapter were diverse ones that varied in their geographical distribution and intensity; nevertheless, it is likely that the reefs at Heron, Lady Musgrave, Fairfax and Green Islands, North Reef and Upolu Cay were significantly degraded as a result.

Notes

1. Report Ref. 70/4028 G.D.Q., D. A. Robinson, 'Heron Island erosion', 20 August 1970, SRS5416/1 Box 10 Item 61, NP231, Bunker – Heron Island, QSA, p2.
2. In-letter, Hon. John Herbert, Minister for Labour and Industry, Brisbane to Hon. H. Richter, Minister for Local Government and Conservation, 13 September 1965, SRS5416/1 Box 10 Item 61, NP231, Bunker – Heron Island, QSA.
3. In-letter Ref. 19.102. A. J. Peel, Director, QDHM, Brisbane to Under-Treasurer, Queensland Treasury, Brisbane, 22 November 1965, SRS5416/1 Box 10 Item 61, NP231, Bunker – Heron Island, QSA.
4. Report Ref. 70/4028 G.D.Q., D. A. Robinson, 'Heron Island erosion', 20 August 1970, SRS5416/1 Box 10 Item 61, NP231, Bunker – Heron Island, QSA, p2.
5. P. Mather, Honorary Secretary, GBRC, 'Statement on the possible effect following construction of a landing strip on Heron Island (Statement compiled by the GBRC)', 22 July 1971, SRS5416/1 Box 10 Item 60, NP 268, Bunker, QSA, p5.
6. OHC 4, 14 January 2003; OHC 9, 28 February 2003; OHC 11, 1 July 2003; OHC 30, 3 October 2003; OHC 44, 4 December 2003.
7. OHC 9, 28 February 2003.
8. OHC 22, 12 September 2003; OHC 28, 19 September 2003; Anonymous, 'Recollections of the reef', Changes in the Great Barrier Reef since European Settlement, Oral History Collection, School of TESAG, JCU, September 2003.
9. OHC 22, 12 September 2003.
10. Out-letter Ref. Res. 6181, Secretary to Surveyor and Property Officer, Queensland Department of the Interior, Brisbane, 28 April 1949, SRS5416/1 Box 10 Item 58, NP220, Bunker, QSA.
11. Cited in *The Sunday Mail*, 26 July 1953, SRS5416/1 Box 10 Item 58, NP220, Bunker, QSA.
12. C. Roff, 'Visit to area in which Naval and Fleet Air Arm exercises are conducted off the Queensland coast', 6 October 1953, SRS5416/1 Box 10 Item 58, NP220, Bunker, QSA, p2 (the aerial photographs mentioned in that report were not found at the QSA).
13. In-letter, Mr D. Jolly to Secretary, Queensland Forestry Department, 20 April 1954, SRS5416/1 Box 10 Item 58, NP220, Bunker, QSA.
14. Cited in *Courier-Mail*, 24 August 1963, found in SRS5416/1 Box 10 Item 58, NP220, Bunker, QSA; W. G. Douglas, Regional Controller, 'N.Q. No. 8 of 1965, Queensland-Hervey Bay, Gunnery and bombing area ND (R699) and Northward', Queensland Department of Shipping and Transport, Brisbane, SRS5416/1 Box 10 Item 58, NP220, Bunker, QSA.
15. In-letter, E. McKeown, National Parks Ranger, Tully to Secretary, 4 October 1938, SRS5416/1 Box 66 Item 447, NP836, Trinity 'R' – Green Island, QSA.
16. Queensland Conservation Council, 'Submission to the QNPWS on the Green Island Management Plan', 1978, SRS5416/1 Box 63 Item 431, NP836, Trinity 'B' Transfer Batch 1, QSA, p4.
17. In-letter, N. H. Traves, Indooroopilly to Director, QNPWS, 31 July 1978, SRS5416/1 Box 63 Item 431, NP836, Trinity 'B' Transfer Batch 1, QSA, p1.
18. OHC 2, 9 November 2002; OHC 11, 1 July 2003; OHC 30, 3 October 2003; OHC 32, 6 October 2003.
19. Town Clerk, Cairns to Chairman, Marine Board Office, Brisbane, 26 September 1931, SRS146/1 Item 2, Correspondence Subject Files – Permit protecting coral and surrounds of Green Island, QSA; GBRMPA, *Green Island Economic Study: Summary Report*, October 1979, Economic Associates Australia, 1979, SRS5416/1 Box 64 Item 434, NP836, Trinity 'B' Transfer Batch 4, QSA, Appendix A.

200 Other impacts on coral reefs

20 I. Bennett, 'Notes made on a trip on MV *Cape Moreton*, September–December 1969', Manuscripts, MS9348, Papers 1944–2000, Box 6, Folder 7, NLA.
21 W. W. M. McCulloch, Superintendent, Yarrabah Mission, Cairns to Chief Protector of Aboriginals, South Brisbane, 21 December 1925, SRS4356/1 Box 27 Item 6818, Aboriginals – Reserves N – Turtle Bay, QSA.
22 In-letter, A. Busuttin, Brampton Island to Land Commissioner, Mackay, 1 October 1935, SRS5416/1 Box 35, SRS5416/1 Box 38 Item 241, NP488, Ingot – Brampton Island, QSA, p3.
23 In-letter Ref. 38/14007B, H. F. Baker, West Bundaberg to Secretary, Queensland Forestry Sub-Department, Brisbane, 29 October 1938, SRS5416/1 Box 9 Item 57, NP224, Bunker – Lady Musgrave Island, QSA.
24 In-letter Ref. 38/14394, E. O. Marks, Honorary Secretary, GBRC, Brisbane to the Hon. F. A. Cooper, Treasurer, Brisbane, 12 January 1938, PRV8340/1 Item 1, QSA.
25 Memo, 'Protection of Islands – Barrier Reef, Circular No. 727', C. J. Trist, Secretary, Queensland Sub-Department of Forestry, Brisbane, 23 March 1939, PRV8340/1 Item 1, QSA.
26 In-letter Ref. 40.2373.11, Secretary, Office of the Commissioner for Railways, Brisbane to Secretary, QDHM, Brisbane, 22 May 1940, PRV8340/1 Item 1, QSA; Out-letter Ref. 225/45, G. Gentry to Secretary, Queensland Sub-Department of Forestry, Brisbane 11 October 1940, PRV8340/1 Item 1, QSA.
27 GBRMPA, *Green Island Economic Study: Summary Report*, October 1979, Economic Associates Australia, 1979, SRS5416/1 Box 64 Item 434, NP836, Trinity 'B' Transfer Batch 4, QSA, Appendix A.
28 Out-letter, E. A. Ferguson, Director, Queensland Tourist Services to Under-Secretary, Queensland Department of Health and Home Affairs, Brisbane, 3 November 1948, RSI920/1 Item 14, General correspondence batches, QSA, pp1–2.
29 In-letter Ref. NPR R382 Dunkalli, National Parks Ranger McKeown, Tully to Secretary, Queensland Department of Forestry, Brisbane, 14 February 1950, SRS5416/1 Box 23 Item 131, NP382, Dunkalli, QSA.
30 OHC 25, 16 September 2003; ORAL TRC3178, Interview with Syd Curtis, December 1994–June 1995, NLA, p166.
31 In-letter, E. McKeown, National Parks Ranger, Tully to Secretary, NP836, Trinity 'P' – Green Island – Underwater Observation Chamber, QSA.
32 GBRMPA, *Green Island Economic Study: Summary Report*, October 1979, Economic Associates Australia, 1979, SRS5416/1 Box 64 Item 434, NP836, Trinity 'B' Transfer Batch 4, QSA, Appendix A.
33 Memo, Director, Queensland Tourist Services to Under-Secretary, Queensland Department of Health and Home Affairs, Brisbane, 3 July 1950, RSI920/1 Item 12, General correspondence batches, QSA.
34 ORAL TRC3178, Interview with Syd Curtis, December 1994–June 1995, NLA, pp91–2.
35 H. S. Curtis, Research Biologist to Secretary, Forestry Department, Brisbane, 12 December 1966, SRS5416/1 Box 9 Item 57, NP224, Bunker – Lady Musgrave Island, QSA; Hon. V. B. Sullivan, Minister for Lands, Brisbane to the Hon. John Herbert, Queensland Minister for Labour and Tourism, Brisbane, 14 November 1969, SRS5416/1 Box 9 Item 57, NP224, Bunker – Lady Musgrave Island, QSA; ORAL TRC3178, Interview with Syd Curtis, December 1994–June 1995, NLA, p166.
36 G. Goeden, Queensland Fisheries Service, Green Island Management Plan: submission on marine resources, 1978, SRS5416/1 Box 63 Item 431, NP836, Trinity 'B' Transfer Batch 1, QSA.
37 I. Bennett, 'Audio-visual: the Great Barrier Reef', Manuscripts, MS9348, Papers 1944–2000, Box 6, Folder 18, Miscellaneous, 1967–1995, NLA, p12.

Chapter 13
Changes in island biota

Introduction

Many of the islands and cays of the Great Barrier Reef have been significantly modified by human activity since European settlement (Hopley, 1982; Lucas et al., 1997, p50). As outlined in previous chapters, those impacts include the removal of guano and rock phosphate (Chapter 10), and the effects of military target practice exercises involving the bombing and shelling of islands and reefs (Chapter 12). In addition to those impacts, however, there have been other profound transformations of the biota of islands, which form the subject of this chapter. Those transformations include the creation of coconut palm plantations on many islands; the destruction of island vegetation by introduced goats; the misuse of fire; and the introduction of exotic types of vegetation, such as *Lantana spp.* and prickly pear (*Opuntia spp.*). Other significant changes in island vegetation and animal populations occurred due to the development of various types of infrastructure on islands, including tourist resorts and airstrips. Those changes are discussed in turn in this chapter, based on evidence gathered primarily from the archival files of the QEPA, held at the QSA.

The creation of coconut palm plantations, 1892–1900

Significant transformations of many islands of the Great Barrier Reef occurred as a result of the introduction of coconut palms, prior to 1900. The establishment of coconut palm plantations took place, with the support of the Queensland government, for several reasons: as a source of employment; as a resource for use by shipwrecked mariners; and to produce copra for export. Consequently, from 1892 to 1900, more than 46 locations were planted with coconut palms, and hundreds of thousands of individual plants were established on islands. The practice of planting coconut palms also occurred in Torres Strait; Reclus (1882, p364) stated that parts of Murray Island, for instance, were also 'clothed with a continuous forest of cocoanut [sic] palms, trees which all travellers assure us were not found in Australia before the arrival of the European immigrants'. His account suggests that the formation of coconut palm plantations on islands accompanied European colonisation and, prior to 1900, may have been a widespread practice.

The creation of coconut palm plantations on the islands of the Great Barrier Reef dates to at least 1892, when the QDAS requisitioned the cutter, *Lizzie Jardine*, for that purpose. The Annual Report of the QDAS for that year stated that:

> Planting has been carried on continuously. The system adopted is to first plant the nuts in nurseries, and at the proper time remove them to a permanent position. [...] Some 6700 nuts have been planted out on the islands lying between Mackay and Sir Charles Hardy Island, to the north of Cooktown, and from the latest reports the early plantations are making vigorous growth.
> (MacLean, 1892, p604)

The same report indicated that, by 1892, 6,747 palms had been planted at 52 island locations, some of which received large numbers of plants: 900 palms were planted on South Palm Island, and 1,064 plants were introduced at 'M' island, between Mackay and the Whitsunday Group. Other types of plantation were also created, including mangoes, guavas and Kauri pine (MacLean, 1892, p604).

By the end of 1894, a total of 8,184 palms had been planted on islands of the Great Barrier Reef: 6,984 of those were established near Mackay, and the remaining 1,200 near Bowen (Blakeney, 1895, p1091). In the annual report of the QDAS for the year 1894–1895, MacLean (1895, p1019) stated:

> The accessible islands in the vicinity of Mackay having been planted, [...] the base of operations has now been changed to Bowen to deal with the islands in that neighbourhood. The plantations already formed are doing well, no disease or vermin having attacked the young palms. During the past year fresh plantations have been made on Seaforth Island, at Eimeo, at Kennedy Sound, on Brampton, Goldsmith, Allonby, and Stone Islands, and a commencement has been made near Bowen.

During the following year, 72,000 palms were planted on islands; subsequently, in 1896, 222,696 palms were planted (Blakeney, 1896, p520; 1897, p986). Figure 13.1 illustrates the numbers of palms planted during 1894–1900, and also the exports of coconuts from Queensland in the following decade, when the palms were producing nuts. However, the distribution of plantations was uneven, as the Registrar-General acknowledged: they were concentrated in the northern part of the Great Barrier Reef, where the plants grew more readily (Blakeney, 1897, p969).

In northern Queensland, the islands of the Great Barrier Reef provided suitable habitats for the plantations because of the availability of salt water, seaweed and marine mud. In his annual report for 1895–1896, in which he announced the establishment of 1,000 new plantings, MacLean (1896, p448) stated: 'Periodical dressings of the plants with sea weed and saline mud, or watering with sea water, has a most beneficial effect upon the plants'. The plantations, therefore, required

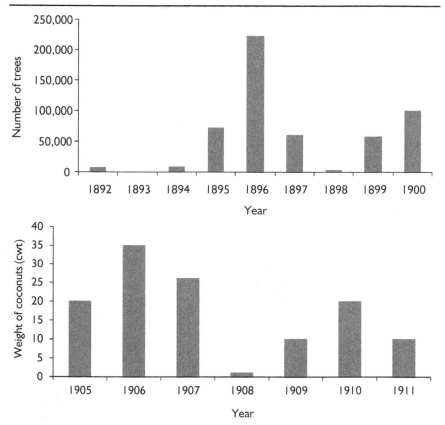

Figure 13.1 (a) Numbers of coconut palms planted in the Great Barrier Reef, 1892–1900; (b) Coconut exports from Queensland, 1905–1911. Source: Author, compiled from data provided in the Annual Reports of the Queensland Registrar-General, QVP, 1895–1900; Annual Reports, QDAS, QVP, 1891–1892; SSQ, 1905–1911

little capital investment once the coconut palms had been transplanted from the nurseries. However, the industry experienced difficulties as a result of damage caused by visitors to the islands. MacLean (1896, p448) stated:

> These islands are from time to time visited by *bêche-de-mer* fishermen and others, who think nothing of leaving a fire burning when they leave an island, and some of them even go so far as to set the grass on fire purposely.

Another source of damage to the early plantations was, reportedly, Indigenous people who visited the islands and who were accused of firing the vegetation and of removing the young plants in order to take the nuts (MacLean, 1897, p912).

By 1897, a large plantation had been formed at Shaw Island, near Kennedy Sound, comprising three smaller plantations (MacLean, 1897, p912). The

204 Changes in island biota

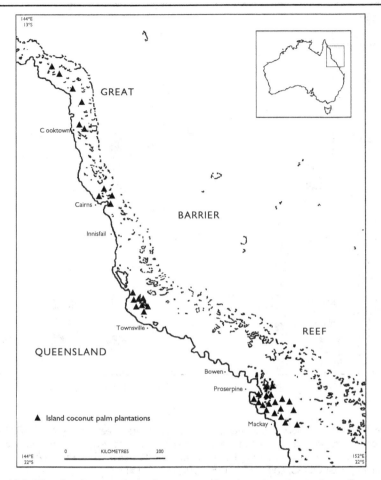

Figure 13.2 The distribution of island coconut palm plantations in northern Queensland, 1898. Source: Author, compiled from data provided in MacLean (1898, p1039)

following year, MacLean (1898, p1039) reported that the prospects for the industry were good, stating:

> In these cocoanut [sic] plantations the foundation of a profitable industry has been laid that could be fostered and built up to the advantage of all concerned without taking into account the great blessing that some of these palms may prove to any shipwrecked crews that may be thrown upon the otherwise rocky and barren islands off our coast.

By the end of 1898, the number of established plantations had reached 72, at separate locations on 47 islands, ranging from the Whitsunday Group to Sir Charles Hardy Island; the distribution of those islands is shown in Figure 13.2.

The total number of plantations increased, and their geographical distribution expanded, since 1892.

In addition to offering a resource for shipwrecked crews, the industry was intended to serve other purposes, both social and economic. MacLean (1898, p1039) stated:

> With regards to the manner in which the conditions of the coast Aborigines [sic] in this district between Bowen and Mackay could be ameliorated, and at the same time ensure the safety and cultivation of the cocoanut [sic] plantations, it is suggested that a station on Carlisle or adjacent land be formed under the care of a married couple. A small steam launch, which would be worked as cheaply as a sailing vessel, to be purchased, and the men employed in tending the present plantations, and as opportunity offered extending their area. As a good many of the palms should fruit next year, the females could be employed in the preparation of copra and fibre.

By 1898, however, the northern plantations had been planted, but not cultivated, so had not yet contributed to the revenue of the colony (Hughes, 1898, p758). By 1900, limited economic benefits had been reaped from the coconut palm plantations, although J. Hughes (1900, p249), the Registrar-General, acknowledged that the plantations were still at that stage very young.

During the next decade, despite the large number of palms planted on Great Barrier Reef islands, the export of coconuts and copra took place on a comparatively small scale in northern Queensland. Figure 13.2 shows the size of the coconut exports from Queensland, from 1905 to 1911, reported in the *SSQ*; the major destinations for the produce were Western Australia, Victoria, South Australia and the Arru Islands. From 1902–1916, copra was also exported from Queensland; the copra trade was greatest in 1906, when 2,904 cwt were exported from Queensland to Victoria. Like the trade in coconuts, however, the Queensland copra export quantities were small in comparison with the produce of some Pacific island states, of which 6,572 cwt were shipped via Queensland in 1907.[1] The comparatively low export quantities suggest that the majority of the coconuts that were grown remained in Queensland, though official reports of the Queensland government do not indicate the destination of the produce. Yet although the Queensland coconut industry was economically insignificant, in comparison with other industries, the plantations represented a significant modification of the landscapes of those islands of the Great Barrier Reef where they were established. Figure 13.3 illustrates the coconut palm plantation at Palm Island: it suggests the scale of the vegetation clearance and modification that was involved in creating a plantation. At many islands, the palms became firmly established as part of the vegetation; numerous reports, written several decades after the palms were planted, describe the survival – and even, at Rabbit Island, the expansion – of the plantations (Clune, 1945, p245).[2]

Figure 13.3 The coconut palm plantation at Palm Island, c.1920. Source: Photograph Ref. 3741 T.B., SRS57/1 Item 26, Queensland Primary Production, Industry, Architecture, Views and People (Photograph Albums), Cairns and District – Barron Falls, Kuranda, Green Is., Atherton Tableland, Malanda, Port Douglas, Queensland State Archives, Brisbane

Overgrazing by introduced goats

Significant destruction of native island vegetation occurred at many islands of the Great Barrier Reef due to the introduction of goats (*Capra hircus*). Many documentary sources and oral history sources provide evidence of the damage to island vegetation caused by goats.[3] Brennan (1988, pp334–5) described the impacts of goats at Brampton, South Percy and North Keppel Islands, acknowledging that the animals altered the vegetation of those islands over a period of eighty years: reducing the cover of grass and heath species and creating scalds. In turn, those impacts exacerbated erosion and facilitated the spread of exotic plant species. Brennan (1988) argued that the impacts of goats were greater on oceanic, rather than continental, islands, because the biota of the former are more susceptible to disturbance. The evidence presented here suggests that, even on continental islands, the presence of goats transformed island landscapes, and native vegetation was able to recover only after the goat populations had been eradicated.

Goats were introduced to Great Barrier Reef islands for several reasons. During the period of early European exploration of the Great Barrier Reef, the animals provided a resource for shipwrecked mariners, as one oral history informant, a zoologist, stated:

> Goats were in many cases left behind by mariners. The early survey vessels and the early guano vessels and whalers – and the guano vessels and whalers tended to be the same vessels – left goats in a number of locations, and coconut trees, for those reasons: that if people got shipwrecked, they had a goat to eat and they had a coconut to provide them with some sort of a liquid.[4]

The same informant stated that goats were also introduced to Lady Elliot Island, from 1863 to 1873, and at Lady Musgrave Island during the 1890s, by lighthouse keepers and guano miners, as a source of milk and meat. Goats were present at Lady Musgrave Island until 1974, when they were eradicated (QNPWS, 1999c).

In addition to Lady Elliot and Lady Musgrave Islands, goats were introduced to many other islands. In 1926, at North Keppel Island, one report stated:

> Mr Walls Senior, a resident of the island, was interviewed. He stated that a small herd of goats existed on the island when he arrived in 1926. Soon after he introduced Sannean goats as a milking herd. He stated that the feral goat herd then began to increase until it stabilized at the present population of approximately 700 to 1,000 goats.[5]

In 1935, goats were found grazing on Digby, Percy, North Palm and Grassy Islands; the latter island, found in the Whitsunday Group, was reported to have at least 600 goats present in 1936 (Birtles, 1935, pp101, 153; Caldwell, 1936, p39).[6] In the same year, large herds of goats were found 'on Lady Musgrave and nearby islands'; the animals were also found on Penrith Island, in the Whitsunday Group, and on the northern part of Long Island. When Boyd Lee left Grassy Island, in 1938, he left behind 'several hundred goats, half a dozen cows, several bulls, a horse, and poultry'; the goats, however, had escaped from their enclosure and were running loose on the island (Caldwell, 1938, p136).[7] Elliot (1950, p9) reported that goat herds were present on Orpheus and North Palm Islands and also that the deliberate transfer of animals between islands occurred:

> Goats and pigs have been raised on Orpheus and have gone wild. Some years ago a batch of goats was transferred from Orpheus to North Palm and these have increased greatly in numbers, much to the satisfaction of the fishermen.

Two other reports, written in 1956 and 1962, indicated that small herds of goats were also present at North Molle and Saddleback Islands.[8]

Documentary evidence describes the impacts of goats on island landscapes. In particular, the destruction of vegetation at Lady Musgrave Island has been documented since 1928, when Napier (1928, pp35–6) provided the following account:

> The undergrowth has been completely eaten away by a flock of goats which have inhabited the place for years, whereas every other island that we saw was clad so thickly in its green and tangled robe of grass and weed and low-hung twisted branches that the crossing of it was a long and hot and complicated task. [...] All the undergrowth has long since gone; hardly a weed can be found from one end of the place to the other; every branch has been denuded of its leaves, and even the bark upon the trees has been gnawed away to the height of several feet.

Another account, written in 1936 by the Queensland Acting Director of Forests, stated that at Lady Musgrave and nearby islands the presence of large herds of goats was threatening both the 'scenic charm' of the islands and the native island wildlife; grazing by goats had resulted in the 'destruction of the grass, herbaceous and shrubby vegetation on the islands [and the] diminution of the normal food for land birds and other Australian animals'.[9]

The destruction of vegetation at Lady Musgrave Island and other islands of the Capricorn-Bunker Group was described in 1936 by a National Parks Ranger, G. Geoffrey, who reported the presence of 'about twenty head of goats' at Lady Musgrave Island, herds of goats in the Bunker Group, and about '150 head of goats' at Fairfax Island, which were reported to be in very poor condition as a result of lack of food and were eating the roots of *Pisonia* trees in order to survive.[10] The herds found in the Bunker Group were also reported by Geoffrey to be in poor condition; he recommended that the animals should be destroyed because 'they are only a means of destroying the vegetation on the islands'.[11] Steers (1938, p55) also described the damage caused by goats at Lady Musgrave Island, although he acknowledged that the damage was less than in previous years. He stated: 'A few years ago there was much less vegetation on Lady Musgrave Island; the change has been brought about by a considerable reduction in the number of goats'. Despite Geoffrey's recommendation that the goats should be destroyed at Lady Musgrave Island, by 1947 the animals continued to destroy the vegetation of the island. Mr Marshall, the Queensland Ichthyologist, stated that 'at present the above island has become over-run with goats. They have eaten the island bare of grass and are now destroying trees'. In 1948, an early attempt was made to eradicate goats from Lady Musgrave Island and a National Parks Ranger reported that all goats on that island were removed and destroyed, although goats were subsequently re-introduced.[12]

Another location that was degraded by goats was Fairfax Island where, in 1953, C. Roff investigated the impacts of military target practice on the island (Chapter 12), and also reported the effects of a herd of about 80 goats. He stated:

> These goats have denuded the island, to the extent that ground-flora could not be found. The only trees on the island are *Pisonia grandis* and one solitary oak, *Casuarina equisetifolia*. The lower foliage of *all* the *Pisonia* trees has been extensively eaten to as far as the goats can reach and climb up the trees. No regeneration in the form of young trees or plants was noticeable on the island, these apparently having been eaten. On the smaller island, goats are not present and this island is covered extensively by both trees and ground-flora. These goats which do not seem to serve any useful purpose are damaging the natural flora on the island.[13]

More than a decade later, in 1965, a National Parks Ranger reported that the *Pisonia* trees of the Fairfax Islands were still being grazed by goats.[14]

Further damage was documented at Lady Musgrave Island during the 1960s, including other impacts besides the destruction of vegetation. In 1964, although the goat population had declined to around 18 animals, many ticks were found on the island.[15] In 1966, Peter Ogilvie, a zoologist at the QNPWS, counted 14 goats at Lady Musgrave Island, which he argued damaged the vegetation and threatened the gull and bridled tern nesting sites on the cay.[16] However, a research biologist with the QNPWS, H. S. Curtis, stated that the situation at Lady Musgrave Island in 1966 – while still unsatisfactory because of the destruction of vegetation – was an improvement since MacGillivray and Rodway had reported the almost complete destruction of the vegetation, in 1927, by around 200–300 goats; nonetheless, Curtis recommended the removal of the remaining goats in order to allow the surviving vegetation to recover.[17]

Oral history evidence provided by Curtis corroborated this account of extensive damage by goats to island vegetation. He reported that 'the devastation caused by goats on coral cays is total', especially at Lady Musgrave Island, which had been 'denuded of vegetation' by the animals. He also stated that:

> Fairfax Island, which is a double island [...], had lots of goats on both of the islands; [on] the larger of the two you wondered how the goats managed to survive because the vegetation was reduced to a group of mature *Pisonia* trees at one end of the island, trimmed flat underneath, up as high as a goat could reach standing on its hind legs. Other than that it was bare coral rubble with here and there a green sprout of a weed or something germinating, but still too far down amongst the coral for a goat to get at it. And I can only think that they managed to eat seaweed at low tide [...].[18]

Curtis also acknowledged that the vegetation of Lindeman Island was severely affected by the large number of goats on that island.

Oral history evidence provided by a zoologist and environmental manager gives a consistent impression of the impacts of overgrazing by goats; however, the damage caused by the animals was exacerbated by lighthouse keepers, who cut trees to provide food for starving goats:

> At some locations, the damage was quite significant. [The goats] systematically wiped out all the vegetation on the island. If you go back through the lighthouse keepers' records, there are records which have these little cryptic comments: 'chopped trees to provide food for goats.' Ultimately, between them, they removed everything from the whole island. On other islands, that didn't have human occupation, where [animals] were left, they basically grazed the island so that it was virtually bare, and they browsed the trees to the extent that they could jump up and get the leaves. I've got photographic evidence from Lady Musgrave and Fairfax Islands where you can look across the whole island and it's almost as if someone's shaved the trees off up to a certain level: there's nice green vegetation above that level, and below it there's nothing at all.[19]

As a result of concerns about this destruction, the removal of goats was advocated from the island National Parks of the Great Barrier Reef.

In 1969, W. Wilkes, the Secretary of the QNPWS, summarised the problem of goats in the Bunker Group National Parks:

> A number of National Park islands under control of this Department are suffering from the ravages of goats but little work has been done on the problem to date. It is clear however that they are a serious threat to the vegetation of an island and as exotic fauna it is desirable to eliminate them from the National Parks. Except in the case of very small islands however, this is difficult to accomplish.[20]

One reason why the eradication of goats was difficult to accomplish was the deliberate transfer of some animals between islands as Julie Booth, a naturalist resident on Fairfax Island, reported in 1969. She stated: 'There have been a large number of campers at Lady Musgrave who visited Fairfax [...]. [They] brought two goats over from the other island, which they were going to leave here'. Consequently, she reported, goats were destroying vegetation at Fairfax Island.[21]

Additional details of the destruction of vegetation at Lady Musgrave Island were sent to the Queensland Government Botanist in 1970, in a report which described the limited spread of *Caesalpinia bonduc* (the native plant known as 'Wait-a-while') on islands where goats were present. The report stated that, although MacGillivray and Rodway recorded the presence of *Caesalpinia bonduc* at Northwest, Hoskyn and Lady Musgrave Islands in 1927, in the *Reports of the Great Barrier Reef Committee*, by 1970 at Lady Musgrave Island only the seeds of that plant were found. That report acknowledged that 'the island was in the final stages of devastation by goats'. In contrast, at Hoskyn Island where no goats were present, *Caesalpinia bonduc* had grown continuously since 1927.[22] At Lady Musgrave Island, the goats remained on the island until their eradication in 1974; elsewhere, the impacts of goats persisted until later, as at North Keppel Island, where in 1975 there remained 800–1,000 goats on the island that were starting to cause erosion on the eastern side of the island by overgrazing.[23]

During the 1970s, in response to substantial evidence of destruction of island vegetation by goats, the systematic eradication of the animals from many islands of the Great Barrier Reef commenced. Feral goats were removed from Lady Musgrave, Fairfax and Hoskyn Islands in 1971. By 1972, about 500 goats had been destroyed at Brampton Island, with around 150 animals remaining on the island, and goats were also removed from Lindeman Island at around the same time.[24] Another major eradication programme occurred at South Percy Island, and by 1976, the removal of goats had also been completed at North Keppel Island.[25] However, the eradication of goats – specifically, the disposal of the carcasses – itself created an environmental problem. Nevertheless, in 1976, a report by J. S. McEvoy, the Senior Zoologist of the QNPWS, about the experimental eradication of goats at North Keppel Island – which contained photographs of the extensive degradation of the eastern foreshore of that island – concluded that eradication programmes, if well organised and conducted with sufficient resources, were an appropriate means of managing the problem of overgrazing by goats.[26]

The impacts of goats on the islands on which they were introduced have been contested. Brennan (1988) argued that the degradation caused by goats occurred primarily to grasses and that trees were comparatively unscathed; he also argued that grass communities at several islands recovered quickly after the eradication of goats was completed. However, Brennan (1988) considered only continental islands, whereas oral history evidence suggests that the cays that supported goat populations were severely degraded by overgrazing.[27] In addition, the documentary evidence presented above suggests that, even at the continental islands, substantial damage was inflicted by goats to many vegetation species, including *Pandanus* and *Pisonia* trees. Furthermore, additional environmental impacts that have been described above – including the import of ticks and the disturbance of gull and bridled tern nesting sites at Lady Musgrave Island – were also attributed to the goat populations. Together, those impacts probably constitute a substantial modification of several island habitats.

Nonetheless, after the eradication of the goats, some recovery of the island vegetation occurred. At Lady Musgrave Island, Curtis stated that when he visited the island in 1966, significant re-growth of vegetation had occurred.[28] The recovery of vegetation at Lady Musgrave Island was described by A. B. Gibb in 1975:

> The removal of goats has led to a marked increase in ground cover plants. Bare shingle edges and ridges of conglomerate exposed by guano mining had been fairly conspicuous features of the cay during the 1969 visit but are now mostly obscured by ground cover plants. There appears to have been a marked increase in the area occupied by the two thickets of *Caesalpinia bonduc*. The smaller patch is at present approx. 90 × 65 ft. and the larger approx. 330 × 215 ft. During the 1969 visit nearly all aerial roots of *Pandanus* had their apices damaged by goats and failed to reach the ground. Aerial roots are now developing normally. Several groups of *Casuarina* seedlings were seen whereas none were noted in 1969. *Casuarina* branches frequently sweep the beach area whereas previously all were trimmed to the maximum height of the goats' reach. In addition to species noted in 1969 the following native or naturalised species were noted in 1975: *Argemore ochroleuca, Boerhavia repens, Euphorbia tannensis, Panicum spp.* [and] *Tournefortia argentea*.[29]

The consequences of the extensive, prolonged damage caused by goats at Lady Musgrave Island are uncertain; no evidence was found to indicate whether vulnerable or rare plant species recovered after the animals had been removed from the cay. The destruction of vegetation by goats, however, may have caused significant changes in the vegetation communities of several islands in the Great Barrier Reef.

Clearance of island vegetation

Destruction of vegetation occurred at some islands for firewood and in order to clear land for planting, as several documentary sources indicate; that activity took place since the period of earliest European settlement. As early as 1857, J. S. V. Mein had cleared the centre of Green Island for firewood and for cultivation. Another island to experience vegetation clearance was Dunk Island where, in 1908, E. J. Banfield (1913, pp12, 43) cleared and burned bloodwood forest in order to prepare land for planting. Later, in 1939, further vegetation clearance occurred at that island as Beach Oak was cut by the crews of trochus luggers in order to supply firewood for their boilers.[30] Clearance of trees on Great Barrier Reef islands also occurred using the method of ringbarking, as V. Grenning, the Queensland Director of Forests, stated in 1951:

> Ringbarking and other destruction on the islands not reserved and given the protection of the National Parks Regulations has already caused some marring of what otherwise must be one of the most beautiful spots in the world.[31]

Another method of clearing island vegetation was using fire; in 1952, B. E. Bailey, the Honorary Secretary of the Magnetic Island United Progress Association, reported that Magnetic Island experienced 'continual burning off by grass and bush fires', causing the destruction of flora, disturbance to fauna, soil erosion, and sediment and nutrient runoff to the fringing reefs of the island.[32]

Further destruction of vegetation was reported in 1954 at Henning and Carlisle Islands, in the Whitsunday Group, where half of the *Casuarina* trees had been destroyed at the northern beach of Henning Island, and other *Casuarina* trees had been cut at Carlisle Island. One report, by F. O. Nixon, stated that at Henning Island 'the wood had been stripped of bark and cut into billets, obviously the work of a crew of a trochus shell boat, several of which are operating in that area'. That destruction was regarded as serious as it would increase the erosion of sand from the beach. Another report of the same year referred to the cutting of *Casuarina* trees, at the northern tip of Henning Island, stating that all parts of the trees that were large enough for firewood were removed.[33] Various sources, therefore, indicate that ringbarking, the use of fire and timber-cutting caused the destruction of vegetation at several islands of the Great Barrier Reef.

Vegetation has also been cleared from numerous islands of the Great Barrier Reef as a result of the development of infrastructure, including tourist resorts, airstrips and research stations. As previously acknowledged (in Chapter 12), the development of infrastructure on islands reflected the expansion of tourism in the Great Barrier Reef, especially since around 1930. The construction of airstrips, which required substantial areas of relatively flat land, was one activity that inevitably degraded island vegetation. An airstrip was constructed on Lindeman Island in July 1946, requiring clearance of the trees. However, by 1957, that airstrip was unserviceable and another was proposed for the island; National

Figure 13.4 The airstrip at Lindeman Island, October 1963. Source: Photograph Ref. 3273, SRS189/1 Box 17 Item 73, Queensland State Archives, Brisbane

Parks Ranger Hausknecht stated that the construction of the new airstrip would involve the destruction of between 1,500 and 2,000 trees. The second airstrip and a terminal building were commissioned in 1957, and air services to the island commenced in April 1958.[34] The Lindeman Island airstrip is shown in Figure 13.4, which illustrates the extent of deforestation that took place. Comparable land clearance took place during the construction of the Dunk Island airstrip, which also involved the drainage of a freshwater swamp to the south of Brammo Bay.[35] Other airstrips were constructed at Brampton Island (although that development also involved land reclamation and modification of part of the island's shoreline), Lady Elliot Island and Lizard Island.[36]

Introduction of exotic species of vegetation

When J. S. V. Mein cleared the centre of Green Island in 1857, that land was used to plant exotic species of vegetation that were obtained from the Botanical Gardens in Sydney, as his account in *The Sydney Morning Herald* indicates:

> The centre of the island we had cleared and planted with a lot of things procured from the Botanical Gardens of Sydney. Among the plants were guava, grape vine, Indian corn, pumpkins, radish, turnips, cabbage, &c.[37]

Mein's alteration of the vegetation composition of Green Island represents an early example of the introduction of exotic species of vegetation to island ecosystems. Many other examples of the introduction of vegetation species are found in documentary records. Before 1931, bananas and pawpaws were planted on Snapper Island, which probably represented the remains of an early Chinese market garden (QEPA, 2003c). In 1933, J. V. Busuttin, the Lessee of the Repulse

Islands, stated: 'I intend to plant fruit trees on Brampton and Goldsmith if the ground is suitable', and in 1935, National Parks Ranger McKeown reported that pawpaws and coconuts had been cultivated at Stephens Island.[38] Birtles (1935, pp182–3), describing Snapper Island, referred to a plantation on the island containing turnips, pawpaws and sweet potatoes.[39]

One banana plantation, at Henning Island in the Whitsunday Group, had a significant impact on the landscape of that island, as National Parks Ranger McKeown indicated. Prior to 1934, Henning Island had been a 'mostly jungle-covered island', but the Licensee of the Island, Dr J. Macdonald, cleared 5 acres of the rainforest on the western slopes, facing the Whitsunday Passage, and established a banana plantation there. By 1936, the bananas were in 'a very dirty and neglected state', and the report stated that the clearing 'has to a large extent spoilt the scenic value of this beautiful island'. By 1938, the banana plants were still in existence at Henning Island, and McKeown suggested that borers had been found at the plantation; consequently, in 1938, the eradication of banana plants on the island commenced.[40] A later investigation of Henning Island, in 1961, after the island was gazetted as a National Park, found that the area that had previously been used for the banana plantation had not yet recovered from that disturbance.[41]

One of the exotic plants introduced to islands that became a particular source of concern was *Lantana spp.*, which by 1935 was established at Brampton Island. A. Busuttin, the Lessee of that island, reported:

> There is a considerable quantity of *Lantana* scattered over the island in places here and there, which appears to have gone beyond the stage where it can be economically cleared by manual labour.[42]

Between 1938 and 1940, *Lantana* was found on several other islands in the Whitsunday Group: Lindeman, Haslewood, Henning and South Molle Islands. At Haslewood Island, the *Lantana* infestation was reported to be 'very heavy over a large area in the vicinity of that once covered by improvements', while at South Molle Island 'a good deal of scattered *Lantana*' was found.[43]

During the 1940s, other exotic species were introduced to islands of the Great Barrier Reef. One invasive species was the prickly pear (*Opuntia spp.*), which was found at Hinchinbrook Island in 1941 and 1946, and also at Masthead Island. In 1947, the cultivation of pine and coconut trees in Mausoleum and Acacia National Parks by the Lessee of Mausoleum Island was referred to the Queensland Sub-Department of Forestry in an attempt to restrict the spread of pests to island habitats (Flood, 1977, p3).[44] However, island residents required a source of food and some of the land found on islands was suitable for cultivation; at Magnetic Island in 1950, for example, National Parks Ranger McKeown stated:

> Good sized areas of flat cultivable land run back from several of the largest bays. These areas have been alienated, and are under cultivation, producing pineapples, mangoes and other tropical fruits.[45]

The cultivation of islands, therefore, represented a conflict between the wish of National Parks officials to conserve native island habitats and the desire of settlers for productive land.

Another infestation of prickly pear was reported in 1951, at Pioneer Point in the Whitsunday Group, and in 1953 a diesoline flame-thrower was used in an attempt to control weeds at Long Island, in the Molle Group. In contrast, other exotic species were introduced deliberately, as at Long Island in 1956, when reportedly severe sand erosion at the island's beaches prompted the introduction of marram grass to stabilise the sand.[46] By 1960, the control of some exotic species on islands had become a significant problem. At Brampton Island, a helicopter was used to spray pesticide in an attempt to destroy *Lantana*, which had heavily infested the island, as the District Forester at Mackay described: 'The *Lantana* has spread right through the bush and presents an impenetrable thicket in places'.[47] In the same year, the Lessees of North Keppel Island were required – as conditions of Special Lease 10756 – to clear the island of 'noxious weeds, noxious plants, *Lantana* and prickly-pear within six months'.[48]

The effects of the introduction of exotic vegetation species to some island ecosystems interacted in various ways with other human activities. For example, P. H. Anderson, the Lessee of Lady Musgrave Island, complained in 1966 that he had planted pawpaw and coconut trees on the island and those had flourished until someone brought goats to the island, which destroyed the plantations.[49] At Goold Island, in 1968, the landscape was altered as a result of interactions between deforestation and the spread of weed species; the District Forester at Atherton reported that at Goold Island introduced weeds had been present for more than 50 years, but they were spreading increasingly rapidly as a result of forest clearance.[50] In one report, A. S. Thorsborne, a naturalist, stated:

> Introduced weeds are taking over [at Goold Island] because so many trees have been and still are being cut down to supply tent poles and firewood. […] The sun beats down on the cleared spots and the native plants and shrubs die – in places there is now no vegetation at all and burr and other noxious weeds infest the vegetated parts.[51]

This report indicates that the direct impacts of vegetation clearance were followed by ecological succession once the pre-existing vegetation had been removed.

Other references to introduced plant species are found in documentary sources: for example, potatoes and onions planted by campers were found at Lady Musgrave Island in 1975; rubbervine was growing at Holbourne Island in 1998; several exotic species colonised Hinchinbrook Island; and *Lantana* and 'sensitive plant' grow at Dunk Island (QNPWS, 1998a, 1999b; QPWS 2000).[52] Overall, the evidence suggests that some significant changes in island vegetation have occurred due to the introduction of exotic plant species. However, further investigation of the vegetation histories of those islands is required if those changes are to be reconstructed in greater detail.

Changes in island fauna

Changes in island fauna since European settlement can be categorised in two groups: changes relating to the destruction of animals – particularly birds – on islands; and transformations of islands resulting from the introduction of exotic species of animals. Many documentary accounts describe the destruction of birds on islands, especially as a result of shooting and the collection of eggs. In January 1944, Mr A. M. Lewis wrote to the QSPCA describing his observations of 'people blazing away with a gun at sea-birds' at Heron Island; another report, written in 1950, referred to parties destroying nesting birds at Wheeler Island, in the Family Group. At Purtaboi Island, in the Family Group, G. S. Stynes of the Dunk Island Tropical Tourist Board reported that 'shooting has been, and is still taking place on Purtaboi Island [and] many terns have met an untimely death at the hands of shooters'.[53] In 1969, K. McArthur, the Honorary Secretary of the Caloundra Branch of the Wildlife Preservation Society of Queensland, stated: 'As is usual at this time of year, we have had reports of pigeon shooting around Hinchinbrook Island National Park'. Shooting of birds, including two nesting white-breasted sea eagles, was reported at Fairfax Island in 1969 by Julie Booth, and A. S. Thorsborne provided evidence of the shooting of 200 Torres Strait pigeons at North Brook Island in 1971. In 1974, A. W. Carle, the Director of the Cape York Environmental Centre, acknowledged that the shooting of large numbers of Torres Strait pigeons at North Brook Island demonstrated that adequate protection of birds had not been enforced.[54]

Other impacts resulted from the collection of birds' eggs from nesting sites by visitors to several islands. Stynes claimed, in 1958, that visitors to Dunk Island were camping, lighting fires and removing hundreds of tern eggs from nests.[55] Concern about the destruction of birds and their rookeries at Lady Musgrave Island was expressed in 1966, and campers from Lady Musgrave also visited Fairfax Island in 1969, where P. H. Anderson reported that 'they collected eggs of the boobies that nested at both islands; another party disturbed the nests, with the result that hundreds of eggs and chicks were taken by gulls'.[56] At Purtaboi Island, in 1972, A. Chisholm wrote that picnickers were disturbing birds by intruding into their nesting sites.[57] In addition to the disturbance of eggs and nesting sites, however, some visitors inflicted deliberate cruelty on birds, as a report by Booth in 1966 demonstrated for Fairfax and Heron Islands:

> You may have noticed a slide or two of a gull's legs tightly bound with wire and strong thread. Some monster goes to their main rookery each season and performs this cruel act of tying the legs up, probably before they can fly. Eventually the legs are so badly damaged that the feet drop off. I cared for six of these birds two years ago at Heron Island (they do not breed there) and another four last year. I have found more here this year.[58]

In addition, at Heron Island in 1970, an observer reported that a 'member of the staff of Heron Island Pty Ltd, Mr Pat Ryan, has over a period of weeks been killing the small birds called White Eyes'.[59]

Some documentary evidence indicates that, as a result of those impacts, bird numbers declined at several islands. In 1970, the Secretary of the QNPWS stated: 'Comparison of the present situation with early records indicates that the tourist resort on Heron Island has resulted in that island being largely deserted by some species of ground-nesting terns'. In 1978, at Michaelmas Cay – one of the most significant sea-bird rookeries in the Great Barrier Reef – the Secretary of the NQNC stated: 'Reports from local and visiting ornithologists reveal that the number of birds present has decreased markedly compared with film and reports taken some few years ago'. Some of the disturbances of seabird colonies at Michaelmas Cay were attributed to the operations of a seaplane taxiing to the beach and conducting low-flying over the cay.[60] Nevertheless, the evidence given above suggests that large numbers of birds had been destroyed, at several islands, before the formation of the GBRMP in 1975.

Other changes in island fauna occurred with the introduction of exotic animal species. In common with the introduction of goats, described above, the transfer of a range of other exotic species of fauna to islands occurred throughout most of the period of European settlement. The animal species introduced were diverse; an indication of that diversity was given by Heatwole (1984, pp28, 31), who reported the presence of feral cats, wallabies, pea fowl, guinea fowl, dogs and an emu on the cays of the Capricornia Section. In some cases, the transformation of island landscapes by introduced animals was prolonged; the QNPWS (1996) reported that sheep grazing commenced in the mid-1800s at the Keppel Islands and continued until the early 1960s, by which time the landscape of the island had been severely degraded. Although some of the introductions of exotic animals were deliberate, others occurred accidentally as animals were carried to islands in ship or boat cargoes; one example of the latter impact occurred at North West Island, where the fowls found on the island in 1928 were reported to have descended from several birds that escaped from a shipwreck (Napier, 1928, p117).

Evidence of the deliberate stocking of islands for pastoralism dates to 1933, when J. V. Busuttin, the Lessee of St Bees Island, reported that his father placed stock on the Repulse Islands, although the animals were later stolen; Busuttin also reported that his brothers had stocked Penrith Island. By 1938, cows, bulls, a horse and poultry had been introduced at Grassy Island by Boyd Lee; horses, sheep and pigs were found at Haslewood Island, in the Whitsunday Group; Mr H. G. Lamond, the Lessee of several Molle Islands, had introduced grazing herds at Molle, Mid Molle, Denman and Planton Islands; and Mr C. F. Pike, the Lessee of Partridge Island, had applied to graze sheep and milking cows on North Molle Island (Caldwell, 1938, p136).[61] In 1948, a National Parks Ranger reported that at Long Island Mr Rasmussen had introduced some cattle and a horse, which were damaging the track on the island. By 1950, eight head of cattle were being grazed on Long Island and over 30 head of stock were present at Lindeman Island, and by 1956 cows had been introduced at North Molle Island and 22 pigs were kept on South Molle Island. By 1960, the Secretary of the QNPWS stated that 'any track work on Brampton Island has been deferred because of the number of horses, sheep and goats on this island'.[62]

In contrast to those reports of the deliberate stocking of islands with grazing animals, other documentary and oral sources indicate that the introduction of pest species also occurred. In 1932, when Cristian Poulsen established the first tourist resort at Heron Island, the cay was 'so infested with rats that it was almost impossible for anything or anybody to live there'.[63] Subsequently, domestic cats were introduced in an attempt to control the rats; however, the cats were allowed to breed and were subsequently reported to be feeding on the juvenile mutton birds on that island. By 1966, rats had also infested other islands, including Fairfax and Wreck Islands, where they were reportedly responsible for significant destruction of the seabird population, as one oral history informant – a zoologist and environmental manager – acknowledged. By 1969, Booth reported that the Fairfax Islands were infested with two species of rats, and by cockroaches.[64] Other exotic fauna included rabbits, which were found by fishermen in 1951 at Humpy Island, near Keppel Island; taipans, which were introduced to South Molle Island in 1957 by Mr Sacks; a fox that also inhabited South Molle Island; an agile wallaby at Heron Island; and African guinea fowl that were released at Lady Musgrave Island in 1974 (Gunn, 1966).[65]

After the formation of the GBRMP, an evaluation of changes in animal populations at Green Island was made in 1978 by Limpus, who concluded that the fauna had 'almost certainly changed from its original state' as a result of the introduction of house sparrows, rats, cats and a small range of reptiles.[66] Other, more recent faunal changes include the presence of cane toads at Lizard Island by 1980; the introduction of feral pigs at Hinchinbrook Island, reported in 1999; and the establishment of feral pigs and cane toads at Dunk Island by 2000 (QNPWS, 1999b, 2000).[67] Those introductions represent the most recent of a series of changes in island fauna that commenced in the earliest period of European settlement in the Great Barrier Reef. As a result of the introduction of a variety of animal species, over an extended period of time, the landscapes of several islands have been substantially modified.

Summary

Various changes in island biota have been described in this chapter: the establishment of coconut palm plantations; overgrazing by introduced goats; the clearance of native vegetation; the destruction of island animal populations (particularly birds); and the introduction of exotic species of plants and animals. Although it is difficult to reconstruct the ecological impacts of those changes, the evidence presented in this chapter suggests that substantial changes must have occurred in many island ecosystems. The biota of some islands – such as Lady Musgrave Island – was significantly altered as a result of multiple impacts (such as guano mining, overgrazing by goats and the construction of tourist facilities), some of which were both prolonged and intensive. Some changes in island biota

– whilst significant – were apparently reversible; at several islands, for instance, some native vegetation degraded by goats apparently recovered after the animals were eradicated (although it is not clear whether ecological communities regained their pre-impacted condition). However, other changes, such as the introduction of *Lantana spp.*, may be much more difficult to reverse, and the coconut palms first established in the late nineteenth century remain a feature of many island landscapes in the Great Barrier Reef today.

Notes

1 SSQ, 1907, p264.
2 In-letter, A. Busuttin, Brampton Island to Land Commissioner, Mackay, 1 October 1935, NP488 Ingot – Brampton Island, QSA; Memo Ref. GJS/CRB, District Forester, Mackay to Secretary, 18 January 1963, SRS5416/1 Box 51 Item 334, NP548, Ossa – Rabbit Island, QSA; In-letter Ref. 2A/NGR, S. K. Robinson, Rabbit Island to Secretary, Land Administration Board, 8 February 1938, SRS5416/1 Box 51 Item 334, NP548, Ossa – Rabbit Island, QSA, p1.
3 For example, OHC 41, 12 November 2003; ORAL TRC3178, Interview with Syd Curtis, December 1994 – June 1995, NLA.
4 OHC 41, 12 November 2003.
5 Cited in In-letter, J. McEvoy, Senior Zoologist, QNPWS to Acting Director, Research and Planning Branch, 28 November 1975, SRS5416/1 Box 38 Item 243, NP612, Keppel – North Keppel Island, QSA, p1.
6 Out-letter Ref. 2A/CT 225/35, Acting Director of Forests to Chairman, Land Administration Board, 1935, SRS5416/1 Box 7 Item 44, NP509-518, Beverley – Beverley Group of Islands, QSA, p2.
7 Out-letter Ref. 895(3) 225/2, A. H. C., Acting Director of Forests to Chairman, Land Administration Board, Brisbane, 18 December 1936, SRS5416/1 Box 10 Item 58, NP220, Bunker, QSA; Memo Ref. 11579 NPR, E. McKeown, NP Ranger, Tully to Secretary, Queensland Forestry Sub-Department, Brisbane, 17 December 1936, SRS5416/1 Box 44 Item 279, NP541-542 Marton, QSA, p2; In-letter, E. McKeown, National Parks Ranger, Tully to Secretary, Queensland Forestry Department, c.1940, SRS5416/1 Box 45 Item 286, NP242, Molle 'A', QSA.
8 Memo Ref. 62/10334, G. J. Swartz, Forester to District Forester, Mackay, 3 September 1962, SRS5416/1 Box 29 Item 187, NP261, Gloucester – Saddleback Island, QSA; In-letter, NP Ranger, Mackay to Secretary, Queensland Department of Forestry, 27 April 1956, SRS5416/1 Box 14 Item 88, NP254, Conway, QSA.
9 Out-letter Ref. Batch 895(3) 225/2, A. H. C., Acting Director of Forests to Chairman, Land Administration Board, Brisbane, 18 December 1936, SRS5416/1 Box 10 Item 58, NP220, Bunker, QSA.
10 In-letter Ref. 225/2, G. Geoffrey, National Parks Ranger to Secretary, Queensland Forestry Sub-Department, Brisbane, 6 October 1936, SRS5416/1 Box 10 Item 58, NP220, Bunker, QSA, pp1–4.
11 In-letter Ref 225/2, A. H. C., Acting Director of Forests to Chairman, Land Administration Board, Brisbane, 17 November 1936, SRS5416/1 Box 10 Item 58, NP220, Bunker, QSA.
12 Out-letter Ref. 47/20202M (Lands) 2A/MC, A. G. M. to V. B. H., 30 June 1947, SRS5416/1 Box 9 Item 57, NP224, Bunker – Lady Musgrave Island, QSA; Out-letter Ref. 4A/MB, Secretary to Forester E. Lister, Bundaberg, 18 October 1948, SRS5416/1 Box 9 Item 57, NP224, Bunker – Lady Musgrave Island, QSA.

220 Changes in island biota

13 C. Roff, 'Visit to area in which Naval and Fleet Air Arm exercises are conducted off the Queensland coast', 6 October 1953, SRS5416/1 Box 10 Item 58, NP220, Bunker, QSA, emphasis in original.
14 Memo Ref. HWH:DMR, NP Ranger, Fairfax Island to Secretary, 'Bird life on Fairfax Island', 19 October 1965, SRS5416/1 Box 10 Item 58, NP220, Bunker, QSA.
15 P. H. Anderson, Bundaberg to Secretary, Queensland Department of Forestry, Brisbane, 12 November 1964, SRS5416/1 Box 9 Item 57, NP224, Bunker – Lady Musgrave Island, QSA.
16 Report Ref. 66/18994 N.P., P. Ogilvie, Zoologist, QNPWS, 'Inspection of Lady Musgrave Island', 1966, SRS5416/1 Box 9 Item 57, NP224, Bunker – Lady Musgrave Island, QSA, p5.
17 In-letter, H. S. Curtis, Research Biologist to Secretary, 12 December 1966, SRS5416/1 Box 9 Item 57, NP224, Bunker – Lady Musgrave Island, QSA.
18 ORAL TRC3178, Interview with Syd Curtis, December 1994–June 1995, NLA, pp143, 147, 149.
19 OHC 41, 12 November 2003.
20 Out-letter Ref. N/LAM 57/40, NP220, W. Wilkes, Secretary to Librarian, Western Australia Department of Fisheries and Fauna, Perth, 30 October 1969, SRS5416/1 Box 10 Item 58, NP220, Bunker, QSA, pp1–2.
21 In-letter, J. Booth, Fairfax Island to W. Wilkes, Secretary, 30 September 1969, SRS5416/1 Box 10 Item 58, NP220, Bunker, QSA.
22 Out-letter Ref. N:RT, Secretary to the Queensland Government Botanist, Indooroopilly, 7 April 1970, SRS5416/1 Box 9 Item 57, NP224, Bunker – Lady Musgrave Island, QSA.
23 In-letter Ref. SL27052, K. J. Cross, Land Inspector, Rockhampton to Land Commissioner, Rockhampton, 28 September 1975, SRS5416/1 Box 38 Item 243, NP612, Keppel – North Keppel Island, QSA.
24 Out-letter, N. McCloy, Secretary to Mr H. Frauca, Bundaberg, 19 November 1974, SRS5416/1 Box 9 Item 57, NP224, Bunker – Lady Musgrave Island, QSA; OHC 41, 12 November 2003.
25 In-letter, Director, Management and Operations to Mr E. J. Harten, Beach Protection Authority, QDHM, Brisbane, 2 November 1976, SRS5416/1 Box 38 Item 243, NP612, Keppel – North Keppel Island, QSA; ORAL TRC3178, Interview with Syd Curtis, December 1994–June 1995, NLA, p150.
26 J. S. McEvoy, Senior Zoologist to Director, Research and Planning Branch, 'Report on an experimental attempt to eradicate feral goats from North Keppel Island', July 1976, SRS5416/1 Box 38 Item 243, NP612, Keppel – North Keppel Island, QSA.
27 ORAL TRC3178, Interview with Syd Curtis, December 1994–June 1995, NLA, p143.
28 ORAL TRC3178, Interview with Syd Curtis, December 1994–June 1995, NLA, p147.
29 A. B. Gibb, 'Report on visit to Lady Musgrave Island', 7–15 July 1975, SRS5416/1 Box 9 Item 57, NP224, Bunker – Lady Musgrave Island, QSA.
30 J. S. V. Mein, 'A cruise inside the Great Barrier Reef, in 1857, and discovery of a reef and harbour', *The Sydney Morning Herald*, 26 February 1866, obtained from the Cairns Historical Society; In-letter Ref. 39/3705, C. Revitt, Honorary Ranger for Bird, Animal and Plant Life, Dunk Island to National Parks Ranger McKeown, Tully, SRS5416/1 Box 23 Item 131, NP382, Dunkalli, QSA.
31 Out-letter Ref. 1A:KE 225/5, V. Grenning, Director of Forests to Chairman, Land Administration Board, Brisbane, 24 July 1951, SRS5416/1 Box 58 Item 392, NP274, Shaw 'A' – Lindeman Island, QSA.
32 B. E. Bailey, Honorary Secretary, Magnetic Island United Progress Association, Arcadia to Hon. V. C. Gair, Queensland Premier, Brisbane, 30 August 1952, RSI920/1 Item 6, General correspondence batches, QSA.

33 In-letter Ref. 54/6815, F. O. Nixon, Secretary, 'Save the Trees' Campaign, Brisbane to Queensland Forestry Department, 2 July 1954, SRS5416/1 Box 72 Item 479, NP259 and NP 270, Whitsunday – Henning Island – Includes NP270 Whitsunday, QSA; In-letter, G. T. McLean, McLean Luxury Cruises, Mackay to Secretary, Queensland Forestry Department, 23 June 1954, NP259, Whitsunday – Henning Island, QSA.

34 Memo to Secretary, 9 November 1956, NP274, Shaw 'A' – Lindeman Island, QSA; Memo, NP Ranger Hausknecht, Mackay to Secretary, 14 October 1957, NP274, Shaw 'A' – Lindeman Island, QSA, p2; In-letter Ref. 7AMC:CN, L. Nicolson, Manager, Lindeman Island Pty Ltd to Secretary, Land Administration Board, Brisbane, 2 December 1957, NP274, Shaw 'A' – Lindeman Island, QSA.

35 Photograph Ref. A2975, Photograph Ref. 3273, SRS189/1 Box 17 Item 73, QSA; SRS5416/1 Box 24 Item 139, NP418, Dunkalli, QSA.

36 SRS189/1 Box 17 Item 73, QSA; SRS5416/1 Box 38 Item 241, NP488, Ingot – Brampton Island, QSA; In-letter, D. C. Adams, Managing Director, Island Airways Pty Ltd to Secretary, Premier's Department, Brisbane, 21 January 1970, SRS5416/1 Box 9 Item 57, NP224, Bunker – Lady Musgrave Island, QSA; Bennett, 'Audio-visual: the Great Barrier Reef', NLA, p12; SRS5416/1 Box 28 Item 179, NP153, Flattery 'A' – Lizard Island, QSA.

37 J. S. V. Mein, 'A cruise inside the Great Barrier Reef, in 1857, and discovery of a reef and harbour', *The Sydney Morning Herald*, 26 February 1866, obtained from the Cairns Historical Society.

38 In-letter Ref. 2A/AK (33.46081B), J. V. Busuttin, St Bees Island to Secretary, Land Administration Board, Brisbane, 6 October 1933, SRS5416/1 Box 45 Item 289, NP246–247, Molle – Repulse Islands, QSA; Memo, H. S. Curtis to Secretary, 14 November 1967, SRS5416/1 Box 38, Item 236, NP807, Hull – Stephens and Sisters Islands – North Island – South Barnard Group, QSA, p2.

39 Out-letter Ref. 225/39, Director to Chairman, Land Administration Board, Brisbane, 13 August 1937, SRS5416/1 Box 2 Item 10, NP64, Snapper Island, QSA.

40 Report Ref. 2A/AK, O.L.544, 5 May 1938, NP259, Whitsunday – Henning Island, QSA; Report Ref. 2A/CT, O.L.544, Bowen Land Agent's District, c.1938, NP259, Whitsunday – Henning Island, QSA; Out-letter Ref. 2A/NGR, O.L.544 Bowen 225/3, Secretary, Forestry to Secretary, Land Administration Board, 5 November 1936, NP259, Whitsunday – Henning Island, QSA.

41 Memo, H. A. Hausknecht, National Parks Ranger, Mackay to Secretary, Queensland Forestry Department, 12 October 1961, NP259, Whitsunday – Henning Island, QSA, p1.

42 In-letter, A. Busuttin, Brampton Island to Land Commissioner, Mackay, 1 October 1935, SRS5416/1 Box 38 Item 241, NP488, Ingot – Brampton Island, QSA, p3.

43 Report Ref. SL12403 7A:KE, 'Special conditions', SRS5416/1 Box 58 Item 392, NP274, Shaw 'A' – Lindeman Island, QSA, p1; Memo Ref. 225/3, E. McKeown, National Parks Ranger, Tully to Secretary, 6 December 1938, SRS5416/1 Box 72 Item 482, NP269, Whitsunday 'A' – Haslewood Island and Surrounds, QSA, p. 2; In-letter Ref. 2A/LC, J. Bergin, Land Commissioner, Mackay to Secretary, Land Administration Board, Brisbane, 30 August 1939, NP259, Whitsunday – Henning Island, QSA; In-letter Ref. SL6792 225/5, J. Bergin, Land Commissioner, Mackay to Secretary, Land Administration Board, Brisbane, 16 October 1940, SRS5416/1 Box 15 Item 90, NP275, Conway 'A', QSA, p2.

44 Memo Ref. WJB/LL, J. R. Dawson, District Forester, Tully to District Forester, Atherton, 7 August 1941, SRS5416/1 Box 35 Item 218, NP95, Hecate 'A', QSA; Memo Ref. 2A/EM, E. McKeown, National Parks Ranger to Secretary, 12 June 1946, SRS5416/1 Box 35 Item 218, NP95, Hecate 'A', QSA; Out-letter Ref. 2A/MC Res. 6505, Secretary

to Mr G. P. McDonald, 28 May 1947, SRS5416/1 Box 51 Item 332, NP540, Ossa – Mausoleum Island, QSA.
45 In-letter Ref. 4876, E. McKeown, National Parks Ranger, Tully to Secretary, Queensland Forestry Department, 9 May 1950, SRS5416/1 Box 42 Item 265, NP456, Magnetic 'A', QSA, p1.
46 In-letter, National Parks Ranger to Secretary, 18 June 1951, SRS5416/1 Box 13 Item 82, NP227, Conway, QSA; In-letter, E. McKeown, National Parks Ranger, Tully to Secretary, Queensland Forestry Department, 16 June 1953, SRS5416/1 Box 45 Item 287, NP242, Molle 'D', QSA; In-letter Ref. TB/AP, Forest Ranger, Tully to Secretary, Queensland Sub-Department of Forestry, 10 February 1956, SRS5416/1 Box 45 Item 287, NP242, Molle 'D', QSA.
47 Memo Ref. JDM:CRB, District Forester, Mackay to Secretary, Queensland Department of Forestry, 1 December 1960, SRS5416/1 Box 38 Item 241, NP488, Ingot – Brampton Island, QSA.
48 'Conditions of Special Lease 10756', 5 May 1960, SRS5416/1 Box 38 Item 243, NP612, Keppel – North Keppel Island, QSA.
49 P. H. Anderson, Bundaberg to Secretary, Queensland Department of Forestry, Brisbane, 1 November 1966, SRS5416/1 Box 9 Item 57, NP224, Bunker – Lady Musgrave Island, QSA.
50 Memo Ref. AWG:LMP, District Forester, Atherton to Secretary, Queensland Department of Forestry, Brisbane, 15 November 1968, SRS5416/1 Box 36 Item 223, NP389, Hecate – Goold Island, QSA, p2.
51 In-letter Ref. 1A:HG, A. S. Thorsborne to Secretary, Queensland Department of Forestry, Brisbane, 20 October 1968, SRS5416/1 Box 36 Item 223, NP389, Hecate – Goold Island, QSA, p2.
52 'Report on visit to Lady Musgrave Island by A. B. Gibb, 7–15 July 1975', SRS5416/1 Box 9 Item 57, NP224, Bunker – Lady Musgrave Island, QSA.
53 Cited in In-letter Ref. 44/2011 G.D.Q., A. E. Cole, Director to Secretary, QDHM, Brisbane, 28 January 1944, SRS5416/1 Box 10 Item 61, NP231, Bunker – Heron Island, QSA; Report Ref. 7A:MB, Innisfail Land Agent's Office, Innisfail, 24 February 1950, SRS5416/1 Box 24 Item 134, NP384, Dunkalli, QSA; In-letter Ref. 7A:CN 75/1, G. S. Stynes, Dunk Island Tropical Tourist Board to Secretary, Forestry, 30 December 1958, SRS5416/1 Box 24 Item 139, NP418, Dunkalli, QSA.
54 In-letter, K. McArthur, Honorary Secretary, Wildlife Preservation Society of Queensland (Caloundra Branch) to Hon. H. Richter, Minister for Conservation, Brisbane, 16 January 1969, SRS5416/1 Box 36 Item 223, NP389, Hecate – Goold Island, QSA; In-letter, J. Booth, Fairfax Island to W. Wilkes, Secretary, Queensland Department of Forestry, Brisbane, 20 March 1969, SRS5416/1 Box 10 Item 58, NP220, Bunker, QSA; In-letter, A. S. Thorsborne to Secretary, Queensland Department of Forestry, 1 February 1971, SRS5416/1 Box 36 Item 223, NP389, Hecate – Goold Island, QSA; In-letter Ref L.6540, A. W. Carle, Director, Cape York Environmental Centre, Gordonvale to Mr H. S. Curtis, Director, Forestry, 7 January 1974, SRS5416/1 Box 36 Item 223, NP389, Hecate – Goold Island, QSA.
55 In-letter Ref. 7A:CN 75/1, G. S. Stynes, Dunk Island Tropical Tourist Board to Secretary, Forestry, 30 December 1958, SRS5416/1 Box 24 Item 139, NP418, Dunkalli, QSA.
56 In-letter, P. H. Anderson, Bundaberg to Secretary, Queensland Department of Forestry, Brisbane, 1 November 1966, SRS5416/1 Box 9 Item 57, NP224, Bunker – Lady Musgrave Island, QSA; In-letter, J. Booth, Fairfax Island to W. Wilkes, Secretary, Queensland Department of Forestry, Brisbane, 30 September 1969, SRS5416/1 Box 10 Item 58, NP220, Bunker, QSA.
57 In-letter, A. Chisholm to Mr C. Haley, Conservator of Forests, Brisbane, 21 October 1972, SRS5416/1 Box 24 Item 139, NP418, Dunkalli, QSA.

58 In-letter, J. Booth, Fairfax Island to Mr W. Wilkes, Secretary, Queensland Department of Forestry, Brisbane, 5 November 1966, SRS5416/1 Box 9 Item 57, NP224, Bunker – Lady Musgrave Island, QSA.
59 In-letter to Director General, QDPI, Brisbane, 28 August 1970, SRS5416/1 Box 9 Item 57, NP224, Bunker – Lady Musgrave Island, QSA.
60 Out-letter, Secretary, QNPWS to Secretary, Land Administration Commission, Brisbane, 23 February 1970, SRS5416/1 Box 9 Item 57, NP224, Bunker – Lady Musgrave Island, QSA, pp1–2, p2; In-letter Ref. 53.1183, Secretary, NQNC, Cairns to Hon. Mr T. Newberry, Minister for National Parks, Recreation and Culture, Brisbane, 12 August 1978, SRS5416/1 Box 62 Item 430, NP779, Trinity 'A', QSA.
61 In-letter Ref. 2A/AK (33.46081B), J. V. Busuttin, St Bees Island to Secretary, Land Administration Board, Brisbane, 6 October 1933, SRS5416/1 Box 45 Item 289, NP246-247, Molle – Repulse Islands, QSA; Memo Ref. 225/3, E. McKeown, National Parks Ranger, Tully to Secretary, 6 December 1938, SRS5416/1 Box 72 Item 482, NP269, Whitsunday 'A', QSA, p2; Memo Ref. 2A/PM 225/5, Secretary, Queensland Sub-Department of Forestry, Brisbane, 'Proposed National Park over Molle, Mid Molle, Denman and Planton Islands', 24 May 1938, SRS5416/1 Box 14 Item 89, NP257, Conway, QSA; In-letter Ref. 2A/PM S.L. 6763, C. F. Pike, Partridge Island to Secretary for Lands, Lands and Survey Office, Brisbane, 25 September 1938, SRS5416/1 Box 14 Item 89, NP257, Conway, QSA.
62 Memo, National Parks Ranger, Mackay to Secretary, Queensland Forestry Department, 11 November 1948, SRS5416/1 Box 45 Item 287, NP242, Molle 'D', QSA; Memo, Hausknecht, National Parks Ranger, Mackay to Secretary, Queensland Forestry Department, 21 February 1950, SRS5416/1 Box 45 Item 287, NP242, Molle 'D', QSA; In-letter Ref. 56/6792, A. A. Fielding, Overseer, South Molle Island to National Parks Ranger, Mackay, 25 August 1956, SRS5416/1 Box 15 Item 90, NP275, Conway 'A', QSA.
63 In-letter Ref. SL7291, C. Poulsen, Heron Island to Secretary, Land Administration Board, Brisbane, 15 May 1944, SRS5416/1 Box 10 Item 61, NP231, Bunker – Heron Island, QSA, p3.
64 In-letter Ref. 44/2011 G.D.Q., A. E. Cole, Director to Secretary, QDHM, Brisbane, 28 January 1944, SRS5416/1 Box 10 Item 61, NP231, Bunker – Heron Island, QSA; In-letter Ref. CN/NC, C. Poulsen, Heron Island to Secretary, Queensland Sub-Department of Forestry, Brisbane, 15 May 1944, SRS5416/1 Box 10 Item 61, NP231, Bunker – Heron Island, QSA, p1; Report Ref. 66/18994 N.P., P. Ogilvie, Zoologist, QNPWS, 'Inspection of Lady Musgrave Island', 1966, SRS5416/1 Box 9 Item 57, NP224, Bunker – Lady Musgrave Island, QSA, pp5–6; OHC 41, 12 November 2003; J. Booth, Fairfax Island to W. Wilkes, Secretary, 20 March 1969, SRS5416/1 Box 10 Item 58, NP220, Bunker, QSA.
65 Out-letter Ref. 51/13260 Lands, Secretary to Land Administration Board, Dept. of Public Lands, Brisbane, 26 November 1951, SRS5416/1 Box 39 Item 250, NP621-627, Keppel, QSA; Out-letter, Secretary to Mr O. M. Bauer, South Molle Island Tourist Resort, 8 August 1957, SRS5416/1 Box 15 Item 90, NP275 Conway 'A', QSA; Memo Ref. 1A/KO, Secretary to National Parks Ranger Hausknecht, Forests Office, Mackay, 19 February 1957, SRS5416/1 Box 15 Item 90, NP275, Conway 'A', QSA; Out-letter, Secretary to Editor, *News Mail*, Bundaberg, 18 November 1974, SRS5416/1 Box 9 Item 57, NP224, Bunker – Lady Musgrave Island, QSA; Out-letter, Director, Fauna Conservation Branch, QDPI, Brisbane to Mr H. Frauca, Bundaberg, 16 January 1975, SRS5416/1 Box 9 Item 57, NP224, Bunker – Lady Musgrave Island, QSA.
66 C. J. Limpus, Green Island Management Plan – Resource Evaluation, 19 September 1978, SRS5416/1 Box 63 Item 431, NP836, Trinity 'B' Transfer Batch 1, QSA, p1.
67 OHC 11, 1 July 2003.

Chapter 14
Conclusion

A multitude of human activities and impacts

This book provides an environmental history of the coral reefs, islands and marine wildlife of the Great Barrier Reef. In doing so, it presents evidence of a multitude of human activities in, and impacts on, that ecosystem. Not all of the important activities and impacts that have affected the Great Barrier Reef are covered here; some – such as human impacts on fish populations, as well as the effects of shipping, dredging and port development in the region – require further research to document their significance. Nevertheless, it has been possible to reconstruct a wide range of activities and impacts, based on the use of documentary and oral sources. Many of those activities and impacts involved the over-exploitation of living resources. The historical fisheries for *bêche-de-mer*, pearl-shell and trochus, some of which dated from at least 1827, caused sustained and intensive impacts on those marine resources, particularly during their early, unregulated periods of operation (Chapter 5). As a result of over-harvesting, reports of the scarcity of *bêche-de-mer* were made as early as 1890; by 1908, the severe depletion of *bêche-de-mer* and pearl oysters had been recognised, and restrictions of those fisheries became necessary. However, continued fishing for those resources meant that, by 1950, the *bêche-de-mer*, pearl-shell and trochus resources of the Great Barrier Reef were almost certainly significantly degraded from their status prior to European settlement, and those resources may not yet have recovered.

Sadly, if predictably, a similar pattern characterised the other historical fisheries in the Great Barrier Reef. Severe impacts on marine turtles – especially hawksbill and green turtles – were sustained due to the operation of the tortoise-shell industry (which had commenced by the 1860s) and the commercial green turtle fisheries (particularly during the 1920s and 1930s); those impacts were exacerbated by the effects of turtle-riding, turtle farming (during the 1970s) and Indigenous hunting of turtles (Chapter 6). Similarly, dugongs were exploited in Queensland waters since 1847 by commercial fisheries (which resulted in the reported local scarcity of the animals by 1888), and their effects were compounded by additional fishing to supply dugong oil to Indigenous settlements, as well as by Indigenous hunting of dugongs (Chapter 7). East Australian humpback whale numbers were substantially reduced by commercial whaling (between 1952 and 1962), leading to the near-collapse of that population, despite the existence of government regulation and scientific monitoring of the industry (Chapter 8). Corals and shells have also been over-harvested at many locations in the Great Barrier Reef due to the activities of both commercial and recreational collectors, with the result that some reefs have been gradually, yet relentlessly, depleted (Chapter 9).

Exploitation of non-living parts of the Great Barrier Reef has also led to some profound transformations of the reefs, islands and cays of the ecosystem. Some islands – including Raine, Lady Elliot, Lady Musgrave, Fairfax and North West Islands – are probably among the most dramatically modified environments of the Great Barrier Reef, not least because of the activities of guano miners who had removed all of the commercially-viable guano from most of those cays by 1900. Other substantial changes occurred at Raine Island (with quarrying and the construction of a navigation beacon, prior to the guano mining); at Holbourne Island (due to rock phosphate mining); and at Upolu and Oyster Cays (as a result of guano mining which persisted until at least 1940) (Chapter 10). Therefore, the extraction of mineral resources from the Great Barrier Reef spanned almost a century and affected at least ten islands; those impacts ranged from the removal of thousands of tons of guano (at their most benign) to the complete alteration of the geomorphology of the cays (at their most extreme). In the cases of Raine and Lady Elliot Islands, the impacts of guano mining remain in the landscape today and may be irreversible. At Lady Musgrave Island, those landscape modifications became visible when as a result of another impact – overgrazing by introduced goats – the surface of that cay was exposed. Coral itself – both living and dead – was also exploited by the coral mining industry in at least twelve reef locations between 1900 and 1940 in order to produce agricultural and industrial lime; that industry removed thousands of tons of coral from some reefs, particularly at Snapper Island, the North Barnard Islands and Kings Reef (Chapter 11).

Further dramatic changes occurred to reefs and islands as a result of blast fishing using dynamite; the construction of access tracks and channels across and through reefs; the use of reefs and islands for military target practice (as at Lady Musgrave and Fairfax Islands); the practice of reef-walking; and the construction of infrastructure, including tourist resorts (Chapter 12). Although they are difficult to reconstruct, some significant changes in island biota have also occurred as a result of various human activities. The establishment of coconut palm plantations on at least 46 islands, between 1892 and 1900, involved the planting of around 500,000 coconut palms, particularly in the Mackay, Townsville, Cairns and Cooktown areas. Further significant changes in island vegetation occurred with the introduction of goats to many islands – some of which held over 1,000 animals – that, in some cases, destroyed all of the grass and shrub vegetation and also affected *Pisonia* trees. Particular impacts due to overgrazing by goats were sustained at Lady Musgrave and Fairfax Islands, both of which were reportedly denuded of vegetation so that bare coral rubble was exposed. Island vegetation was also destroyed in other ways – such as by the misuse of fire – and many changes in island vegetation have occurred due to the introduction of exotic species, such as *Lantana spp.* and *Opuntia spp.* Similarly, the animal populations of many islands have been altered by both the destruction of native animals (particularly birds) and the introduction of exotic species of animal (Chapter 13).

Unprecedented exploitation of the Great Barrier Reef

Taken together, the environmental changes reconstructed in this book form a narrative of sustained, extensive damage to some coral reefs, islands and organisms of the Great Barrier Reef as a result of the over-exploitation of resources and the degradation and destruction of habitats. There is unequivocal evidence (both documentary and oral) that some parts of the Great Barrier Reef have experienced severe impacts – that have varied in their location and intensity – since European settlement. Consequently, although the Great Barrier Reef remains one of the best-protected coral reef ecosystems in the world, some of its habitats were clearly far from pristine at the time of the formation of the GBRMP, in 1975. In large part due to human activities, some coral reef areas have been degraded to the extent that recovery to their former condition is now unlikely, and many other reef areas have been affected to a lesser extent.

Similarly, there is abundant evidence that some islands of the Great Barrier Reef – particularly Raine, Lady Elliot, Lady Musgrave, Fairfax, North West and Holbourne Islands – were subjected to considerable exploitation and had been significantly modified by the time the GBRMP was established. Moreover, some of the impacts sustained by marine wildlife species in the Great Barrier Reef have been severe and, whilst their ecological consequences may be difficult to establish, the over-exploitation of some populations and species has substantially depleted their numbers and increased their vulnerability to other environmental changes. Overall, the evidence presented in this book suggests that the Great Barrier Reef was exploited earlier, for a longer period, in more locations and more intensively than has previously been documented. The multiple human impacts on the Great Barrier Reef that have occurred since European settlement amount to an unprecedented period of exploitation of, and change in, the ecosystem. In particular, this account illustrates the damage that can be inflicted on coral reefs and their associated habitats and populations in the absence of effective regulation, monitoring and management. Yet, in many cases, the degradation of the Great Barrier Reef due to human activities continued for many years – and sometimes for many decades – after the signs of decline were first recognised, and even after regulations and restrictions of the most destructive practices were introduced.

What is the relevance of a historical account of human activities and impacts – many of which have now ceased (or at least are regulated) in the Great Barrier Reef? This account can inform contemporary management of the ecosystem in several ways. First, environmental histories of this type can potentially be used to identify ecological baselines against which subsequent environmental changes may be assessed. Second, this account provides evidence of some activities (such as coral mining) in the Great Barrier Reef that were previously little-known, if not entirely forgotten, yet which have nevertheless left imprints on the landscapes of some reefs and islands. Third, this book contains salutary lessons, such as the fact that unsustainable industries (such as the commercial humpback whale fishery based at Tangalooma) can very rapidly – and perhaps catastrophically – deplete marine populations, even in the presence of regulation and scientific monitoring of the industry. Yet the scale and intensity of the depletion that probably characterised other, unregulated and unmonitored fisheries, especially during the early period of European exploitation of the Great Barrier Reef, might have eclipsed even those managed fisheries, with unknown ecological consequences. Therefore, effective, scientific management of reef resources is critical if the World Heritage values of the Great Barrier Reef are to be protected, and for some marine species there may now be no sustainable harvest. Fourth, the evidence assembled in this book suggests that the cumulative, localised impacts of many relatively small activities (such as coral and shell collecting), over many decades, was probably not negligible; on the contrary, they probably transformed some coral reef areas, albeit in ways that were imperceptible to contemporary observers. Hence this account provides evidence in support of the concept of shifting baselines, where successive generations of observers fail to notice that an environment or ecosystem is gradually changing, and each generation adopts a new sense of what represents normality for that environment or ecosystem. Many documentary and oral history sources attest to the problem of shifting baselines; my research showed that, without the use of scientific methods, only very few privileged observers – typically those with access to detailed records spanning several years (if not decades) – were able to detect environmental changes reliably in the Great Barrier Reef.

Finally, this environmental history suggests that, whilst individual human impacts (such as military target practice at the Fairfax Islands) can be unequivocally destructive, in many cases it is the interplay of multiple activities and impacts that has created particular outcomes for the reefs, islands and organisms of the Great Barrier Reef. Thus, one of the earliest European impacts in the Great Barrier Reef – the construction of the navigation beacon at Raine Island in 1844, together with the subsequent guano mining at that cay – continues, as an unintended consequence, to cause environmental problems today. This is because the mining and quarrying activities created small cliffs that now present hazards to the green turtles that come ashore on the island to nest. The problem is particularly acute because the cay is the largest and most important green turtle rookery in the world; it therefore plays a critical role in maintaining the population of this vulnerable species. Moreover, the problem of additional turtle mortality at Raine Island is compounded by the increasingly frequent flooding of green turtle nests by seawater, driven at least in part by rising sea level, which results in the mortality of the embryos. Hence, at this remote cay, the effects of two historical human activities (quarrying and guano mining) continue to interact with the biology and ecology of a long-lived, slow-reproducing, vulnerable marine species – one that has also sustained a wide range of other human impacts, including commercial fishing – in the context of ongoing, critical changes in its nesting habitat.

This example illustrates the fact that historical impacts on environments and ecosystems such as the Great Barrier Reef do not necessarily end when the original human activity ceases; they may continue to influence subsequent environmental changes, perhaps irrevocably. Overall, the unprecedented historical impacts on, and exploitation of, the Great Barrier Reef reconstructed in this book must have increased the vulnerability of its reefs, islands and marine wildlife to a range of other stresses. In the context of growing concerns about the future of the Great Barrier Reef – and the urgent need to increase its resilience to natural and human pressures, including the effects of climate change and ocean acidification – the ecosystem now requires an unprecedented level of research, monitoring and effective protection if it is to adapt to the changes that lie ahead.

References

Agassiz, A. (1898) *A Visit to the Great Barrier Reef of Australia in the Steamer 'Croydon,' During April and May, 1896*, Museum of Comparative Zoology, Cambridge, MA.

Agassiz, G. R. (ed.) (1913) *Letters and Recollections of Alexander Agassiz*, Boston, Houghton Mifflin.

Allen, B. and Montell, W. L. (1981) *From Memory to History: Using Oral Sources in Local Historical Research*, The American Association for State and Local History, Nashville, TN.

Allen, G. M. (1942) *Extinct and Vanishing Mammals of the Western Hemisphere, with the Marine Species of All the Oceans*, American Committee for International Wildlife Protection, New York.

Almond, T. M. (1899) Annual Report, Queensland Marine Dept., 1898–1899, QVP, vol. 3, pp1025–1032.

Anonymous (1861) 'Dugong oil: correspondence', *The Lancet*, vol. 78, issue 1995, p508.

Anonymous (1929) *Lindeman Island Tourist Resort, Whitsunday Passage, North Queensland*, A. D. Nicolson, Lindeman Island.

Anonymous ('Viator') (1932) 'The Great Barrier Reef: the world's greatest coral banks', *Cummins and Campbell's Monthly Magazine*, vol. 5, no 57, pp76–77.

Anonymous (1933a) 'Annual Returns of the Townsville Harbour Board', *Cummins and Campbell's Monthly Magazine*, vol. 5, no 71, p37.

Anonymous (1933b) 'Piscatorial: big fish in Townsville Harbour', *Cummins and Campbell's Monthly Magazine*, vol. 5, no 74, pp33–35.

Anonymous ('Investigator') (1937) 'The Great Barrier Reef: one of the world's greatest wonders', *Cummins and Campbell's Monthly Magazine*, vol. 5, no 124, pp38–39.

Antill, E. (1952) *Death on the Barrier Reef*, Hammond, London.

Archer, M., Burnley, I., Dodson, J., Harding, R., Head, L. and Murphy, P. (1998) *From Plesiosaurs to People: 100 Million Years of Australian Environmental History*, Australia: State of the Environment Technical Paper Series (Portrait of Australia), Department of the Environment, Canberra.

Arthington, A. H., Marshall, J. C., Rayment, G. E., Hunter, H. M. and Bunn, S. E. (1997) 'Potential impacts of sugarcane production on riparian and freshwater environments', in Keating, B. A. and Wilson, J. R. (eds), *Intensive Sugarcane Production: Meeting the Challenges Beyond 2000*, CAB International, Wallingford, Oxfordshire, pp403–421.

Baglin D. and Mullins, B. (1969) *Australia's Great Barrier Reef: Wonderland of Coral Cays and Rocky Isles, Fantastic Marine Life and Tropical Vegetation*, Horwitz Publications Inc., North Sydney.

Ball, M. S. and Smith, G. W. H. (1992) *Analysing Visual Data*, Qualitative Research Methods, vol. 24, Sage Publications, Newbury Park, CA.

Banfield, E. J. (1908) *The Confessions of a Beachcomber: Scenes and Incidents in the Career of an Unprofessional Beachcomber in Tropical Queensland*, T. Fisher Unwin, London.

Banfield, E. J. (1913) *The Confessions of a Beachcomber: Scenes and Incidents in the Career of an Unprofessional Beachcomber in Queensland*, 3rd edition, T. Fisher Unwin, London.

Barrett, C. (1930) 'The Great Barrier Reef and its isles: the wonder and mystery of Australia's world-famous geographical feature', *National Geographic Magazine*, vol. 58, no 3, pp354–384.

Barrett, C. (1943) *Australia's Coral Realm: Wonders of Sea, Reef and Shore*, Robertson and Mullens, Melbourne.

Bartlett, N. (1947) 'By air to the Reef', *South West Pacific*, New Series no 18, pp6–9.

Bartley, N. (1892) *Opals and Agates; Or, Scenes under the Southern Cross and the Magelhans: Being Memories of Fifty Years of Australia and Polynesia, With Nine Illustrations*, Gordon and Gotch, London.

Bauer, F. H. (1964) 'Historical geography of white settlement in part of northern Australia: Part 2, the Katherine-Darwin region', CSIRO Division of Land Research and Regional Survey, Divisional Report no 64/1, April 1964, CSIRO, Canberra.

Baxter, I. N. (1990) 'Green Island information review', Research Publication no 25, GBRMPA, Townsville.

Beaglehole, J. C. (ed.) (1955) *The Voyage of the Endeavour, 1768–1771*, Cambridge University Press, Cambridge.

Bedford, R. (1928) *The Great Barrier Reef: A Series of Photographs by E. F. Pollock and Frank Hurley*, Art in Australia, Sydney.

Benham, C. (1949) *Diver's Luck: A Story of Pearling Days*, Angus and Roberston, Sydney.

Bennett, G. (1860) *Gatherings of a Naturalist in Australia*, John van Voost, London.

Bennett, G. H. (1897) 'Annual Report, Inspector of Pearl-shell Fisheries', *QVP*, vol. 2, part 2, pp680–683.

Bennett, G. H. (1898) 'Annual Report, Inspector of Pearl-shell Fisheries', *QVP*, vol. 3, pp1041–1048.

Bennett, G. H. (1899) 'Annual Report, Inspector of Pearl-shell Fisheries', *QVP*, vol. 5, part 2, pp994–997.

Bennett, G. H. (1900) 'Annual Report, Inspector of Pearl-shell Fisheries', *QVP*, vol. 3, part 2, pp1317–1321.

Bennett, I. (1971) *The Great Barrier Reef*, Lansdowne Press, Sydney.

Benson, J. S. and Redpath, P. A. (1997) 'The nature of pre-European native vegetation in south-eastern Australia: a critique of Ryan, D. G., Ryan, J. R. and Starr, B. J. (1995) 'The Australian landscape – observations of explorers and early settlers', *Cunninghamia*, vol. 5, no 2, pp285–328.

Bird, E. C. F. (1971) 'The fringing reefs near Yule Point, north Queensland', *Australian Geographical Studies*, vol. 9, no 2, pp107–115.

Birtles, D. (1935) *North-West by North: A Journal of a Voyage*, Jonathan Cape, London.

Birtles, T. G. (1988) 'European interpretation of the Atherton-Evelyn "vine scrub" of tropical north Queensland, 1880–1920', in Dargavel, J., Dixon, K. and Semple, N. (eds), *Changing Tropical Forests: Historical Perspectives on Today's Challenges in Asia, Australasia and Oceania*, Workshop meeting, Canberra, 16–18 May 1988, Centre for Resource and Environmental Studies, Canberra.

Birtles, T. G. (1997) 'First contact: colonial European perceptions of tropical Queensland rainforest and its people', *Journal of Historical Geography*, vol. 23, no 4, pp393–417.

Blakeney, W. T. (1895) Annual Report, Registrar-General, 1894, QVP, vol. 3, part 2, pp1051–1091.

Blakeney, W. T. (1896) Annual Report, Registrar-General, 1895, QVP, vol. 4, pp477–520.

Blakeney, W. T. (1897) Annual Report, Registrar-General, 1896, QVP, vol. 4, part 2, pp941–986.

Bolton, G. C. (1963) *A Thousand Miles Away: A History of North Queensland*, Jacaranda Press, Brisbane.

Bolton, G. C. (1981) *Spoils and Spoilers: Australians Make Their Environment 1788–1980*, Allen and Unwin, Sydney.

Boult, E. J., Stevens, J. H., Morgan, A., Hamilton W. and Beal, G. L. (1913) 'Report of the Treasury Departmental Committee upon the fisheries industry of the State of Queensland (other than pearl-shell) and Appendices', QPP, vol. 3, pp1037–1056.

Bowen, G. (1864) 'Containing reports upon the formation of a new settlement at Cape York, at the northernmost point of the Australian continent; and upon the completion of the survey of the inside of the Great Barrier Reef, off the north-east coast, by Commander Robinson, R.N.', *Proceedings of the Royal Geographical Society of London*, Session 1863–1864, vol. 8, no 4, pp114–118.

Bowen, G. F. (1889) *Thirty Years of Colonial Government: Selection from the Despatches and Letters of the Right Hon. Sir George Ferguson Bowen*, ed. S. Lane-Poole, vol. 1, Longmans, Green and Co., London.

Bowen, J. (1994) 'The Great Barrier Reef: towards conservation and management', in Dovers, S. (ed.), *Australian Environmental History: Essays and Cases*, Oxford University Press, Melbourne, pp234–256.

Bowen, J. and Bowen, M. (2002) *The Great Barrier Reef: History, Science, Heritage*, Cambridge University Press, Cambridge.

Boyd, A. J. (1882) *Queensland*, Queensland Government Emigration Office, London.

Brannen, J. (ed.) (1992) *Mixing Methods: Qualitative and Quantitative Research*, Avebury, Aldershot.

Brennan, P. F. (1988) 'Forest history of the continental islands: Great Barrier Reef', in Dargavel, J., Dixon, K. and Semple, N. (eds), *Changing Tropical Forests: Historical Perspectives on Today's Challenges in Asia, Australasia and Oceania, Workshop Meeting, Canberra, 16–18 May 1988*, Centre for Resource and Environmental Studies, Canberra, pp327–337.

British Museum (Natural History) (1930–1968) *Great Barrier Reef Expedition 1928–1929: Scientific Reports*, 7 volumes, The British Museum, London.

Cairns Harbour Board (1929) 'Annual Report, Fourth Revised Pamphlet on The Port of Cairns, North Queensland, Australia', Cairns Post Pty Ltd, Cairns.

Caldwell, N. (1936) *Fangs of the Sea*, Angus and Robertson, Sydney.

Caldwell, N. (1938) *Titans of the Barrier Reef: Further Adventures of a Shark Fisherman*, Angus and Robertson, Sydney.

Campbell, F. T. (1887) 'Report on the oyster fisheries of Moreton Bay', QVP, vol. 4, pp115–129.

Carruthers, D. S. (1969) 'Limestone mining', in *The Future of the Great Barrier Reef: Papers of an Australian Conservation Foundation Symposium, Sydney, 3 May 1969*, comp. Australian Conservation Foundation, Australian Conservation Foundation, Parkville, pp47–50.

Chesterman, H. G. (1973) *Sailing Directions, Lightship Cape Moreton from South Island (Burnett River) to Torres Strait: Including Coral Sea Lights and Weather Stations, and the Great North East Channel*, Queensland Museum, Brisbane.

Chilvers, B. L., Lawler, I. R., MacKnight, F., Marsh, H., Noad, M. and Paterson, R. (2005) 'Moreton Bay, Queensland, Australia: an example of the co-existence of significant marine mammal populations and large-scale coastal development', *Biological Conservation*, vol. 122, pp559–571.

Choquenot, D. and Bowman, D. M. J. S. (1998) 'Marsupial megafauna, Aborigines and the overkill hypothesis: application of predator-prey models to the question of Pleistocene extinction in Australia', *Global Ecology and Biogeography Letters*, vol. 7, no 7, pp167–180.

Christesen, C. B. (1936) 'Roving the coral seas', *Walkabout*, 1 June 1936, pp28–31.

Clune, F. (1945) *Free and Easy Land*, Angus and Robertson, Sydney.

Commonwealth of Australia Productivity Commission (2003) 'Industries, land use and water quality in the Great Barrier Reef catchment', Research Report, Commonwealth of Australia, Productivity Commission, Canberra.

Commonwealth of Australia Royal Commission on the Pearl-shelling Industry (1913) *Progress Report*, Government Printer, Melbourne.

Corkeron, P. (1997) 'The status of cetaceans in the GBRMP', in Wachenfeld, D. R. et al. (eds), *State of the GBRWHA Workshop: Proceedings of a Technical Workshop Held in Townsville, Queensland, Australia, 27–29 November 1995*, Workshop Series No. 23, Townsville, GBRMPA, pp283–286.

Coulter, E. J. (1952) Annual Report, Chief Inspector of Fisheries, 1951–1952, QPP, vol. 1, part 2, pp1009–1012.

Coulter, E. J. (1953) Annual Report, Chief Inspector of Fisheries, 1952–1953, QPP, vol. 1, part 2, pp1012–1016.

Coulter, E. J. (1954) Annual Report, Chief Inspector of Fisheries, 1953–1954, QPP, vol. 1, part 2, pp1001–1006.

Coulter, E. J. (1956) Annual Report, Chief Inspector of Fisheries, 1955–1956, QPP, vol. 1, part 2, pp1014–1019.

Coulter, E. J. (1957) Annual Report, Chief Inspector of Fisheries, 1956–1957, QPP, vol. 1, 1957, pp1012–1016.

Coulter, E. J. (1958) Annual Report, Chief Inspector of Fisheries, 1957–1958, QPP, vol. 1, 1958, pp1012–1016.

Coulter, E. J. (1959) Annual Report, Chief Inspector of Fisheries, 1958–1959, QPP, vol. 1, 1959, pp1093–1098.

Coulter, E. J. (1960) Annual Report, Chief Inspector of Fisheries, 1959–1960, QPP, vol. 1, 1960, pp1149–1154.

Courtenay, P. P. (1978) 'Agriculture in North Queensland', *Australian Geographical Studies*, vol. 16, no 1, pp29–42.

Cronon, W. (1992) 'A place for stories: nature, history and narrative', *Journal of American History*, vol. 78, no 4, pp1347–1376.

Crossland, C. J., Done, T. J. and Brunskill, G. J. (1997) 'Potential impacts of sugarcane production on the marine environment', in Keating, B. A. and Wilson, J. R. (eds), *Intensive Sugarcane Production: Meeting the Challenges Beyond 2000*, CAB International, Wallingford, Oxfordshire, pp423–436.

Crowley, G. M. and Garnett, S. T. (2000) 'Changing fire management in the pastoral lands of Cape York Peninsula of northeastern Australia, 1623 to 1996', *Australian Geographical Studies*, vol. 38, no 1, pp10–26.

Cumbrae-Stewart, F. W. S. (1930) *The Boundaries of Queensland: With Special Reference to the Maritime Boundary and the 'Territorial Waters Jurisdiction Act, 1878'*, Government Printer, Brisbane.

Dakin, W. J. (1934) *Whaleman Adventures: The Story of Whaling in Australian Waters and Other Southern Seas Related Thereto, From the Days of Sail to Modern Times*, Angus and Robertson, Sydney.

Davitt, M. (1898) *Life and Progress in Australasia*, Methuen and Co., London.

Denton, S. F. (1889) *Incidents of a Collector's Rambles in Australia, New Zealand, and New Guinea*, Lee and Shepard, Boston, MA.

Denzin, N. K. and Lincoln, Y. S. (2000) 'Introduction: the discipline and practice of qualitative research', in Denzin, N. K. and Lincoln, Y. S. (eds), *Handbook of Qualitative Research*, 2nd edition, Sage Publications, Thousand Oaks, CA, pp1–28.

Dey, I. (1993) *Qualitative Data Analysis: A User-Friendly Guide for Social Scientists*, Routledge, London.

Dick, J. D. W. (1930) Annual report, Acting Chief Inspector of Fisheries, 1929–1930, QPP, vol. 2, pp39–41.

Dick, J. D. W. (1931) Annual report, Acting Chief Inspector of Fisheries, 1930–1931, QPP, vol. 2, pp5–6.

Dick, J. D. W. (1932) Annual report, Chief Inspector of Fisheries, 1931–1932, QPP, vol. 2, pp6–7.

Dick, J. D. W. (1935) Annual report, Chief Inspector of Fisheries, 1934–1935, QPP, vol. 1, pp1091–1093.

Dick, J. D. W. (1936) Annual report, Chief Inspector of Fisheries, 1935–1936, QPP, vol. 1, pp1141–1144.

Dick, J. D. W. (1937) Annual report, Chief Inspector of Fisheries, 1936–1937, QPP, vol. 1, pp1403–1404.

Dick, J. D. W. (1938) Annual report, Chief Inspector of Fisheries, 1937–1938, QPP, vol. 1, pp1283–1284.

Dobbs, K. (2001) *Marine Turtles in the GBRWHA: A Compendium of Information and Basis for the Development of Policies and Strategies for the Conservation of Marine Turtles*, GBRMPA, Townsville.

Dodds, D. R. (2004) 'The Cairns Black Marlin industry', Cairns Historical Society Bulletin, no 509, Cairns Historical Society, Cairns.

Domm, S. (1970) 'Corals of the Great Barrier Reef', *Royal Commission on Great Barrier Reef Petroleum Drilling, Transcript of Proceedings, 22 May 1970–14 July 1970*, vol. 1, AGPS, Brisbane, pp44–46.

Dovers, S. (1994) 'Australian environmental history: introduction, review and principles', in Dovers, S. (ed.), *Australian Environmental History: Essays and Cases*, Oxford University Press, Melbourne, pp2–19.

Dovers, S. (2000) 'Still settling Australia: environment, history and policy', in Dovers, S. (ed.), *Environmental History and Policy: Still Settling Australia*, Oxford University Press, South Melbourne, Victoria, pp2–23.

Eden, C. H. (1872) *My Wife and I in Queensland: An Eight Years' Experience in the Above Colony, With Some Account of Polynesian Labour*, Longmans, Green and Co., London.

Elder, G. H., Pavalko, E. K. and Clipp, E. C. (1993) *Working with Archival Data: Studying Lives*, Qualitative Applications in the Social Sciences, no 88, Sage Publications, Newbury Park, CA.

Elliot, P. (1950) *Smoked Mackerel: A Tale of the Coral Sea and the Great Barrier Reef*, Robert Hale Ltd, London.

Ellis, A. F. (1936) *Adventuring in Coral Seas*, Angus and Robertson, Sydney.

Evans, R. (2007) *A History of Queensland*, Cambridge University Press, Cambridge.

Fison, C. S. (1886) 'Report on the oyster fisheries of Moreton Bay, 1885', *QVP*, vol. 3, part 2, pp827–837.

Fison, C. S. (1887) 'Report on the oyster fisheries of Moreton Bay', *QVP*, vol. 4, part 1, pp121–123.

Fison, C. S. (1888) 'Report on the oyster and other industries within the ports of Moreton Bay and Maryborough', *QVP*, vol. 3, part 2, pp761–775.

Fison, C. S. (1889) 'Report on the oyster fisheries of Moreton Bay and Great Sandy Island Strait', *QVP*, vol. 3, part 2, pp937–941.

Fison, C. S. (1894) 'Report on the oyster and other fisheries of Moreton Bay', *QVP*, vol. 3, part 2, pp1112–1113.

Fison, C. S. (1897), Annual Report, Inspector of Fisheries, 1896, *QVP*, vol. 4, pp617–637.

Fison, C. S. (1899) Annual Report, Fisheries, 1898–1899, *QVP*, vol. 3, pp1039–1040.

Fison, D. (1935) Annual Report, QDHM, 1934–35. *QPP*, vol. 1, pp1083–1104.

Fison, D. (1936) Annual Report, QDHM, 1935–36. *QPP*, vol. 1, pp1133–1156.

Fison, D. (1937) Annual Report, QDHM, 1936–37. *QPP*, vol. 1, pp1393–1415.

Fison, D. (1938) Annual Report, QDHM, 1937–38. *QPP*, vol. 1, pp1271–1296.

Fitzgerald, R. (1982) *From The Dreaming to 1915: A History of Queensland*, University of Queensland Press, St. Lucia, Queensland.

Fitzgerald, R. (1984) *From 1915 to the early 1980s: A History of Queensland*, University of Queensland Press, St. Lucia, Queensland.

Flood, P. G. (1977) 'Coral cays of the Capricorn and Bunker Groups, Great Barrier Reef, Australia', *Atoll Research Bulletin*, no 195, pp1–7.

Flood, P. G. (1984) 'Variability of shoreline position on five uninhabited islands of the Capricornia Section, Great Barrier Reef Marine Park', in Ward, W. T. and Saenger, P. (eds), *The Capricornia Section of the Great Barrier Reef: Past, Present and Future*, The Royal Society of Queensland and ACRS, Brisbane, pp17–23.

Forrester, V. (1925) Annual Report, Queensland Marine Department, 1924–1925, *QPP*, vol. 3, pp291–300.

Forrester, V. (1926) Annual Report, Queensland Marine Department, 1925–1926, *QPP*, vol. 2, pp925–934.

Forrester, V. (1927) Annual report, Queensland Marine Department, 1926–27, *QPP*, vol. 2, pp949–958.

Forrester, V. (1928) Annual report, Queensland Marine Department, 1927–28, *QPP*, vol. 1, pp1181–1190.

Forrester, V. (1929) Annual report, Queensland Marine Department, 1928–29, *QPP*, vol. 2, pp949–958.

Foxton, J. F. G. (1898) '*The Land Act, 1897*, and *The Pastoral Leases Act of 1869*', *QVP*, vol. 3, part 2, pp903–906.

Frankfort-Nachmais, C. and Nachmais, D. (1992) *Research Methods in the Social Sciences*, 4th edition, Arnold, London.

Furnas, M. (2003) *Catchments and Corals: Terrestrial Runoff to the Great Barrier Reef*, AIMS, Townsville.

Gammage, B. (1994) 'Sustainable damage: the environment and the future', in Dovers, S. (ed.), *Australian Environmental History: Essays and Cases*, Oxford University Press, Melbourne, pp258–267.

Ganter, R. (1994) *The Pearl-shellers of Torres Strait: Resource Use, Development and Decline, 1860s–1960s*, Melbourne University Press, Carlton, Victoria.

Garside, A. L., Smith, M. A., Chapman, L. S., Hurney, A. P. and Magarey, R. C. (1997) 'The yield plateau in the Australian sugar industry: 1970–1990', in Keating, B. A. and Wilson, J. R. (eds), *Intensive Sugarcane Production: Meeting the Challenges Beyond 2000*, CAB International, Wallingford, Oxfordshire, pp103–124.

GBRMPA (2000) 'Whale and Dolphin Conservation in the GBRMP', Policy Document, April 2000, GBRMPA, Townsville.

GBRMPA (2009) 'Great Barrier Reef General Reference Map, Map No. SDC 04120A3, August 2009', http://www.gbrmpa.gov.au/__data/assets/pdf_file/0003/8166/GBRMPA_Aug_09_General_Reference_A3.pdf, accessed 18 December 2013.

GBRMPA (2013) 'Facts about the Great Barrier Reef', http://www.gbrmpa.gov.au/about-the-reef/facts-about-the-great-barrier-reef, accessed 25 September 2013.

Gill, J. C. H. (1988) *The Missing Coast: Queensland Takes Shape*, The Queensland Museum, Brisbane.

Gillham, B. (2000) *The Research Interview*, Continuum, London.

Glenne, M. (1938) *Great Australasian Mysteries*, Stanley Paul, London.

Goddard, E. J. (1933) 'The economic possibilities of the Great Barrier Reef', in Tilghman, D. C. (ed.), *The Queen State: A Handbook of Queensland Compiled Under Authority of the Government of the State*, John Mills Himself, Brisbane, pp217–222.

Goeden, G. (1978) *Green Island Management Plan: Submission on Marine Resources*, Queensland Fisheries Service, Brisbane.

Golding, W. R. (1979) *Beyond Horizons*, Wholesale Book and Library Suppliers, Brisbane.

Great Barrier Reef Fisheries Ltd (1929) *Tapping the Wealth of the Great Barrier Reef*, Great Barrier Reef Fisheries Ltd, Sydney.

Great Barrier Reef Protection Interdepartmental Committee Science Panel (2003) 'A report on the study of land-sourced pollutants and their impacts on water quality in and adjacent to the Great Barrier Reef', www.premiers.qld.gov.au/ about/reefwater.pdf, accessed 30 January 2003.

Griggs, P. (1997) 'The origins and development of the small cane farming system in Queensland, 1870–1915', *Journal of Historical Geography*, vol. 23, no 1, pp46–61.

Griggs, P. (1999a) 'Alien agriculturalists: non-European small farmers in the Australian sugar industry, 1880–1920', in Ahluwalia, P., Ashcroft, B. and Knight, R. (eds), *White and Deadly: Sugar and Colonialism*, Nova Science Publishers, Commack, NY, pp135–155.

Griggs, P. (1999b) 'Sugar demand and consumption in colonial Australia, 1788–1900', in Dare, R. (ed.), *Food, Power and Community: Essays in the History of Food and Drink*, Wakefield Press, Adelaide, pp74–90.

Griggs, P. (2000) 'Sugar plantations in Queensland, 1864–1912: origins, characteristics, distribution and decline', *Agricultural History*, vol. 74, no 3, pp1–25.

Griggs, P. (2003) 'Australian scientists, sugar cane growers and the search for new gummosis-resistant and sucrose-rich varieties of sugar cane, 1890–1920', *Historical Records of Australian Science*, vol. 14, no 3, pp291–311.

Griggs, P. (2004) 'Improving agricultural practices: science and the Australian sugarcane grower, 1864–1915', *Agricultural History*, vol. 78, no 1, pp1–33.

Griggs, P. D. (2005) 'Environmental change in the sugar cane producing lands of Eastern Australia, 1865–1990', Conference paper presented at the International Congress of Historical Sciences, University of New South Wales, Sydney, 3–7 July 2005.

Griggs, P. D. (2006) 'Soil erosion, scientists and the development of conservation tillage techniques in the Queensland sugar industry, 1935–1995', *Environment and History*, vol. 12, no 3, pp233–268.

Griggs, P. D. (2007) 'Deforestation and sugar cane growing in Eastern Australia, 1860–1995', *Environment and History*, vol. 13, no 3, pp255–283.

Griggs, P. D. (2011) *Global Industry, Local Innovation: The History of Cane Sugar Production in Australia, 1820–1995*, Peter Lang, Bern.

Gunn, J. (1966) *Barrier Reef by Trimaran*, Collins, London.

Haddon, A. C. (1901) *Head-Hunters: Black, White, and Brown*, Methuen, London.

Hakim, C. (1987) *Research Design: Strategies and Choices in the Design of Social Research*, Routledge, London.

Hamilton, J., Dawson, A., Hoolan, J., O'Connell, W. H. B. and Smyth W. (eds) (1897) 'Report, together with minutes of evidence and proceedings of the Commission appointed to inquire into the general working of the laws regulating the pearl-shell and bêche-de-mer fisheries in the Colony, Queensland Departmental Commission on Pearl-shell and Bêche-de-mer Fisheries', *QVP*, vol. 2, part 2, pp1273–1352.

Harriott, V. J. (2001) 'The Sustainability of Queensland's Coral Harvest Fishery', CRC Reef Research Centre Technical Report no 40, CRC Reef Research Centre, Townsville.

Heatwole, H. (1984) 'The cays of the Capricornia Section, Great Barrier Reef Marine Park, and a history of research on their terrestrial biota', in Ward, W. T. and Saenger, P. (eds), *The Capricornia Section of the Great Barrier Reef: Past, Present and Future*, The Royal Society of Queensland and ACRS, Brisbane, pp25–44.

Hedley, C. (1924) 'Australian pearl fisheries', *The Australian Museum Magazine*, vol. 2, no 1, pp5–11.

Hedley, C. (1925) 'The natural destruction of a coral reef', *Reports of the Great Barrier Reef Committee*, vol 1, pp1–28.

Heinsohn, G. E., Lear, R. J., Bryden, M. M., Marsh, H. and Gardner, B. R. (1978) 'Discovery of a large population of dugongs off Brisbane, Australia', *Environmental Conservation*, vol. 5, pp91–92.

Heinsohn, R., Lacy, R. C., Lindenmayer, D. B., Marsh, H., Kwan, D. and Lawler, I. R. (2004) 'Unsustainable harvest of dugongs in Torres Strait and Cape York (Australia) waters: two case studies using population viability analysis', *Animal Conservation*, vol. 7, pp417–425.

Hill, D. (1960) 'The Great Barrier Reef', *Journal of the Geological Society of Australia*, vol. 7, pp412–413.

Hill, R., Baird, A. and Buchanan, D. (1999) 'Aborigines and fire in the Wet Tropics of Queensland, Australia: ecosystem management across cultures', *Society and Natural Resources*, vol. 12, no 3, pp205–223.

Hill, R., Griggs, P. and Bamanga Bubu Ngadimunku Incorporated (2000) 'Rainforests, agriculture and Aboriginal fire-regimes in Wet Tropical Queensland, Australia', *Australian Geographical Studies*, vol. 38, no 2, pp138–157.

Hoegh-Guldberg, O. (1999) 'Climate change, coral bleaching and the future of the world's coral reefs', *Marine and Freshwater Research*, vol. 50, no 8, pp839–866.

Hoegh-Guldberg, O., Mumby, P. J., Hooten, A. J., Steneck, R. S., Greenfield, P., Gomez, E., Harvell, C. D., Sale, P. F., Edwards, A. J., Caldeira, K., Knowlton, N., Eakin, C.

M., Iglesias-Prieto, R., Muthiga, N., Bradbury, R. H., Dubi, A. and Hatziolos, M. E. (2007) 'Coral reefs under rapid climate change and ocean acidification', *Science*, vol. 318, pp1737–1742.

Hoggart, K., Lees, L. and Davies, A. (2002) *Researching Human Geography*, Arnold, London.

Holmes, C. H. (1933) *We Find Australia*, Hutchinson, London.

Holmes, R. (1918) Annual Report, Pearl-shell fisheries, 1917, *QPP*, vol. 1, pp1667–1669.

Holthouse, H. (1976) *Ships in the Coral*, Macmillan, Melbourne.

Hopley, D. (1982) *The Geomorphology of the Great Barrier Reef: Quaternary Development of Coral Reefs*, John Wiley and Sons, New York.

Hopley, D. (1988) 'Anthropogenic influences on Australia's Great Barrier Reef', *Australian Geographer*, vol. 19, no 1, pp26–45.

Hopley, D. (1989) *The Great Barrier Reef: Ecology and Management*, Longman Cheshire, Melbourne.

Hopley, D. (1994) 'Continental shelf reef systems', in Carter, R. W. G. and Woodroffe, C. D. (eds), *Coastal Evolution: Late Quaternary Shoreline Morphodynamics*, Cambridge University Press, Cambridge, pp303–340.

Hopley, D. (1997) 'Natural heritage attribute: geological and geomorphological aspects', in Lucas, P. H. C., Webb, T., Valentine. P. S. and Marsh, H. (1997) *The Outstanding Universal Value of the Great Barrier Reef World Heritage Area*, GBRMPA, Townsville, pp140–144.

Hopley, D. (2009) 'Geomorphology of coral reefs with special reference to the Great Barrier Reef', in Hutchings, P., Kingsford, M. J. and Hoegh-Guldberg, O. (eds), *The Great Barrier Reef: Biology, Environment and Management*, Coral Reefs of the World, vol. 2, Springer, Dordrecht, pp5–16.

Hopley, D., Smithers, S. G. and Parnell, K. E. (eds) (2007) *The Geomorphology of the Great Barrier Reef: Development, Diversity, and Change*, Cambridge University Press, Cambridge.

Horton, D. (ed.) (1994) *The Encyclopaedia of Aboriginal Australia: Aboriginal and Torres Strait Islander History, Society and Culture*, Aboriginal Studies Press for the AIATSIS, Canberra.

Hughes, I. A. (1937) *In the Wake of the 'Cheerio': The Narrative of a Memorable Cruise in Great Barrier Reef Waters*, I. A. Hughes, Sydney.

Hughes, J. (1898) Annual Report, Registrar-General, 1898, *QVP*, vol. 2, pp725–779.

Hughes, J. (1900) Annual Report, Registrar-General, 1900, *QVP*, vol. 4, part 1, pp209–270.

Hutchings, P., Kingsford, M. J. and Hoegh-Guldberg, O. (2009) 'Introduction to the Great Barrier Reef', in Hutchings, P., Kingsford, M. J. and Hoegh-Guldberg, O. (eds), *The Great Barrier Reef: Biology, Environment and Management*, Coral Reefs of the World 2, Springer, Dordrecht, pp1–4.

Huxham, J. (1925) Annual report, Agent-General, 1924–25. *QPP*, vol. 3, pp1–23.

Huxham, J. (1928) Annual report, Agent-General, 1927–28. *QPP*, vol. 2, pp653–680.

IUCN (2013) 'The IUCN Red List of Threatened Species', Version 2013.2, http://www.iucnredlist.org/, accessed 19 January 2014.

Jacobs, J. M. (1996) *Edge of Empire: Postcolonialism and the City*, Routledge, London.

James, G. (1962) 'Turtle hunt', *Walkabout*, March 1962, pp15–17.

Johnson, M. (2002) '"A modified form of whaling": the Moreton Bay dugong fishery, 1846–1920', in Johnson, M. (ed.) *Brisbane: Moreton Bay Matters*, Brisbane History Group Papers No. 19, Brisbane History Group, Brisbane, pp27–38.

Johnson, N. C. (1988), *Memories: Burrum Heads*, The Publication Committee, Burrum Heads.

Johnston, W. R. (1988) *A Documentary History of Queensland*, University of Queensland Press, St Lucia, Queensland.

Jones, D. (1961) *Cardwell Shire Story*, Jacaranda Press, Brisbane.

Jones, D. (1973) *Hurricane Lamps and Blue Umbrellas: A History of the Shire of Johnstone to 1973*, The Cairns Post Pty Ltd, Cairns.

Jones, D. (1976) *Trinity Phoenix: A History of Cairns and District*, The Cairns Post Pty Ltd, Cairns.

Jones, D. (1980) *The Whalers of Tangalooma*, The Nautical Association of Australia, Melbourne.

Jones, D. (2002) 'The whalers of Tangalooma 1952–1962', in Johnson, M. (ed.), *Brisbane: Moreton Bay Matters*, Brisbane History Group Papers, no 19, Brisbane History Group, Brisbane, pp87–94.

Joske, H. D. A. (1930) *A Life to Live*, Popular Publications, South Melbourne.

Jukes, J. B. (1847) *Narrative of the Surveying Voyage of H.M.S. Fly*, vol. 2, T. and W. Boone, London.

Jukes, J. B. (1871) *Letters and Extracts from the Addresses and Occasional Writings of J. Beete Jukes*, ed. C. A. Browne, Chapman and Hall, London.

Keating, B. A., Verburg, K., Huth, N. I. and Robertson, M. J. (1997) 'Nitrogen management in intensive agriculture: sugarcane in Australia', in Keating, B. A. and Wilson, J. R. (eds), *Intensive Sugarcane Production: Meeting the Challenges Beyond 2000*, CAB International, Wallingford, Oxfordshire, pp221–242.

Kerr, J. (1995) *Northern Outpost*, 2nd edition, Mossman Central Mill Co., Mossman.

King, N. J. (1965) 'Soil acidity and liming', in King, N. J. et al. (eds), *Manual of Cane-Growing*, Angus and Robertson, Sydney, pp99–109.

Lack, C. (1968) 'Dugong fishing in early Queensland', *Newsletter of the Royal Australian Historical Society*, November 1968, pp4–6.

Ladd, H. S. (1970) 'Preliminary report on conservation and controlled exploitation of the Great Barrier Reef', *Royal Commission on Great Barrier Reef Petroleum Drilling: transcript of proceedings*, vol. 1, Government Printer, Brisbane, pp53–57.

Lamond, H. G. (1936) 'Sharking', *Walkabout*, 1 March 1936, pp23–25.

Lawrence, D. and Cornelius, J. (1993) 'History, relics and tower graffiti', in Smyth, A. K., Zevering, K. H. and Zevering, C. E. (eds), *Raine Island and Environs Great Barrier Reef: Quest to Preserve a Fragile Outpost of Nature*, Raine Island Corporation and GBRMPA, Brisbane, pp1–11.

Lawrence, D., Kenchington, R. and Woodley, S. (2002) *The Great Barrier Reef: Finding the Right Balance*, Melbourne University Press, Carlton South, Victoria.

Limpus, C. J. (1978) 'The Reef: uncertain land of plenty', in Lavery, H. J. (ed.), *Exploration North: A Natural History of Queensland*, revised edition, Lloyd O'Neill, South Yarra, pp187–222.

Limpus, C. J. (1980) 'The Green turtle, *Chelonia mydas* in eastern Australia', Management of Turtle Resources: Proceedings of a Seminar held jointly by Applied Ecology Pty Ltd and the Department of Tropical Veterinary Science at Townsville, Queensland, Australia, 28 June 1979, JCU Research Monograph 1, JCU, Townsville.

Limpus, C. J. (1997) 'Marine turtles of the Great Barrier Reef World Heritage Area', in Wachenfeld, D. R. et al. (eds), *State of the GBRWHA Workshop: Proceedings of a*

Technical Workshop held in Townsville, Queensland, Australia, 27–29 November 1995, Workshop Series No. 23, GBRMPA, Townsville, pp256–258.

Limpus, C. J., Miller, J. D., Parmenter, C. J. and Limpus, D. J. (2003) 'The green turtle, *Chelonia mydas*, population of Raine Island and the northern Great Barrier Reef: 1843–2001', *Memoirs of the Queensland Museum*, vol. 49, pp349–340.

Linedale, J. C. (1922) 'Warden's report on gold and mineral fields for 1921', QPP, vol. 2, pp599–604.

Loch, I. (1984) 'Raine Island: a foul hen roost', *Australian Natural History*, vol 1, part 5, pp181–183.

Loch, I. (1991) 'Michaelmas Cay', *Cairns Shell News*, vol. 11, no 49, p5.

Loch, I. (1994) 'Raine Island: a foul hen roost', *Australian Natural History*, vol. 1, part 5, pp181–183.

Lock, A. C. C. (1955) *Destination Barrier Reef*, Georgian House, Melbourne.

Loos, N. (1982) *Invasion and Resistance: Aboriginal–European Relations on the North Queensland frontier, 1861–1897*, ANU Press, Canberra.

Lough, J. M. (1999) 'Sea surface temperatures on the Great Barrier Reef: a contribution to the study of coral bleaching', Research Publication no 57, GBRMPA, Townsville.

Love, R. (2000) *Reefscape: Reflections on the Great Barrier Reef*, Allen and Unwin, St Leonard's, New South Wales.

Loyau, G. E. (1897) *The History of Maryborough and Wide Bay and Burnett Districts from the Year 1850 to 1895; Compiled from Authentic Sources*, Pole, Outridge and Co., Brisbane.

Lucas, P. H. C., Webb, T., Valentine. P. S. and Marsh, H. (1997) *The Outstanding Universal Value of the Great Barrier Reef World Heritage Area*, GBRMPA, Townsville.

Lunine, J. (1999) *Earth: Evolution of a Habitable World*, Cambridge University Press, Cambridge.

Lurie, R. (1966) *Under the Great Barrier Reef*, Jarrolds, London.

Mackay, J. (1911) Annual Report, Queensland Marine Department, 1910–1911, QPP, vol. 3, pp1183–1198.

Mackay, J., Douglas, H. A. C. and Bennett, G. H. (1908) *Report, Together with Minutes of Proceedings, Minutes of Evidence Taken before the Commission, and Appendices, Queensland Royal Commission Appointed to Inquire into the Working of the Pearl-shell and Bêche-de-mer Industries*, Government Printer, Brisbane.

Mackellar, C. D. (1912) *Scented Isles and Coral Gardens: Torres Straits, German New Guinea, and the Dutch East Indies*, John Murray, London.

MacKnight, C. C. (1976) *The Voyage to Marege: Macassan Trepangers in Northern Australia*, Melbourne University Press, Carlton, Victoria.

MacLean, P. (1892) Annual Report, QDAS, 1891–1892, QVP, vol. 4, pp604–605.

MacLean, P. (1895) Annual Report, QDAS, 1894–1895, QVP, vol. 3, part 2, pp1015–1025.

MacLean, P. (1896) Annual Report, QDAS, 1895–1896, QVP, vol. 4, pp443–454.

MacLean, P. (1897) Annual Report, QDAS, 1896–1897, QVP, vol. 4, part 2, pp903–916.

MacLean, P. (1898) Annual Report, QDAS, 1897–1898, QVP, vol. 3, part 2, pp1035–1041.

McLoughlin, L. C. (2000) 'Shaping Sydney Harbour: sedimentation, dredging and reclamation 1788–1990s', *Australian Geographer*, vol. 31, no 2, pp183–208.

Marks, H. V. (1933) *A Christmas Holiday on the Great Barrier Reef, 1932–1933*, Harris and Sons, Sydney.

Marsh, H. and Corkeron, P. (1997) 'The status of the dugong in the Great Barrier Reef Marine Park', in Wachenfeld, D. R., Oliver, J. K. and Davis, K. (eds), *State of the Great Barrier Reef World Heritage Area Workshop: Proceedings of a Technical Workshop Held in Townsville, Queensland, Australia, 27–29 November 1995*, Workshop Series no 23, GBRMPA, Townsville, pp234–235.

Marsh, H. and Lawler, I. (2001) *Dugong Distribution and Abundance in the Southern Great Barrier Reef Marine Park and Hervey Bay: Results of an Aerial Survey in October–December 1999*, Final Report, GBRMPA, Townsville.

Marsh, H., O'Shea, T. J. and Reynolds, J. E. III (2011) *Ecology and Conservation of the Sirenia: Dugongs and Manatees*, Cambridge University Press, Cambridge.

Marsh, H., De'ath, G., Gribble, N. and Lane, B. (2005) 'Historical marine population estimates: triggers or targets for conservation? The dugong case study', *Ecological Applications*, vol. 15, pp481–492.

Marsh, H., Penrose, H., Eros, C. and Hugues, J. (2002), *Dugong: Status Report and Action Plans for Countries and Territories*, UNEP, Nairobi.

Marsh, H., Lawler, I., Kwan, D., Delean, S., Pollock, K. and Alldredge, M. (2004) 'Aerial surveys and the potential biological removal technique indicate that the Torres Strait dugong fishery is unsustainable', *Animal Conservation*, vol. 7, pp435–443.

Mass, O. (1975) *Dangerous Waters*, Rigby, Adelaide.

Maxwell, W. G. H. (1968) *Atlas of the Great Barrier Reef*, Elsevier, Amsterdam.

McCarthy, F. D. (1955) 'Aboriginal turtle hunters', *The Australian Museum Magazine*, 15 March 1955, pp283–286.

McCracken, G. (1988) *The Long Interview*, Qualitative Research Methods Series, vol. 13, Sage Publications, Newbury Park, CA.

McLoughlin, L. C. (1999) 'Environmental history, environmental management and the public record: will the records be there when you need them?', *Australian Journal of Environmental Management*, vol. 6, no 4, pp207–218.

Meyer, W. S. (1997) 'The irrigation experience in Australia – lessons for the sugar industry', in Keating, B. A. and Wilson, J. R. (eds), *Intensive Sugarcane Production: Meeting the Challenges Beyond 2000*, CAB International, Wallingford, Oxfordshire, pp437–454.

Moorhouse, F. W. (1935) 'Notes on the Green turtle (*Chelonia mydas*)', *Reports of the GBRC*, vol. 4, pp1–22.

Musgrave, A. and Whitley, G. P. (1926) 'From sea to soup: an account of the turtles of North-West Islet', *The Australian Museum Magazine*, vol. 2, pp331–336.

NADC (1946) *Pearl-Shell, Bêche-de-Mer and Trochus Industry of Northern Australia*, NADC, Maribyrnong, Victoria.

Napier, E. S. (1928) *On the Barrier Reef: Notes from a No-ologist's Pocket-Book*, Angus and Robertson, Sydney.

Noonan, M. (1962) *Flying Doctor on the Great Barrier Reef*, Hodder and Stoughton, London.

Nordstrom, K. F. and Jackson, N. L. (2001) 'Using paintings for problem-solving and teaching physical geography: examples from a course in coastal management', *Journal of Geography*, vol. 100, no 5, pp141–151.

Northman, J. (1933) 'Magnetic Island: novel fishing: larks with sharks', *Cummins and Campbell's Monthly Magazine*, vol. 5, no 74, p39.

Nott, J. F. (2003) 'Intensity of prehistoric tropical cyclones', *Journal of Geophysical Research*, vol. 108, no D7, 4212, pp1–11.

Nott, J. F. and Hayne, M. (2001) 'High frequency of "super-cyclones" along the Great Barrier Reef over the past 5,000 years', *Nature*, vol. 413, pp508–511.
O'Donoghue, J. L. (2001) *Robert Hayles Snr, A Magnetic Life: The Story of His Family and Hayles Magnetic Island Pty Ltd*, J. L. O'Donoghue, Brisbane.
Oliver, J. (1985) 'An Evaluation of the Biological and Economic Aspects of Commercial Coral Collecting in the Great Barrier Reef Region', Report to GBRMPA, GBRMPA, Townsville.
Orams, M. B. and Forestell, P. H. (1994) 'From whale harvesting to whale watching: Tangalooma 30 years on', in Bellwood, O., Choat, H. and Saxena, N. (eds), *Recent Advances in Marine Science and Technology*, '94, JCU, Townsville, pp667–673.
Palmer, B. H. (1879) *The Exhibition Essay on the Geographical Features, Natural Resources and Productions of the Cook District of Queensland*, Cleghorn and Co., Brisbane.
Paterson, R., Paterson, P. and Cato, D. (1994) 'The status of Humpback whales Megaptera novaeanglia in east Australia thirty years after whaling', *Biological Conservation*, vol. 70, pp135–142.
Peel, A. J. (1961) Annual Report, QDHM, 1960–1961, *QPP*, vol. 1, pp745–754.
Peel, A. J. (1962) Annual Report, QDHM, 1961–1962, *QPP*, vol. 1, pp757–764.
Peel, A. J. (1963) Annual Report, QDHM, 1962–1963, *QPP*, vol. 2, pp827–838.
Peel, A. J. (1966) Annual Report, QDHM, 1965–1966, *QPP*, vol. 2, pp841–857.
Powell, J. M. (1988) *An Historical Geography of Modern Australia: The Restive Fringe*, Cambridge University Press, Cambridge.
Powell, J. M. (1994) 'A legacy of competing imperatives – environment and development in Australia since 1788', *Land Degradation and Rehabilitation*, vol. 5, no 2, pp89–106.
Powell, J. M. (1996) 'Historical geography and environmental history: an Australian interface', *Journal of Historical Geography*, vol. 22, no 3, pp253–273.
Preen, A. and Marsh, H. (1995) 'Response of dugongs to large-scale loss of seagrass from Hervey Bay, Queensland, Australia', *Wildlife Research*, vol. 22, pp507–519.
Pulsford, J. S. (1996) *Historical nutrient usage in coastal Queensland river catchments adjacent to the Great Barrier Reef Marine Park*, Research Publication no 40, GBRMPA, Townsville.
Puotinen, M. L., Done, T. J. and Skelly, W. C. (1997) *An Atlas of Tropical Cyclones in the Great Barrier Reef Region: 1969–1997*, CRC Reef Research Centre, Technical Report No. 19, CRC Reef Research Centre, Townsville, pp92–120.
QEPA (1999), *State of the Environment Queensland 1999*, QEPA, Brisbane.
QEPA (2003a) *Green Island National and Marine Park: Park Guide*, QEPA, Brisbane.
QEPA (2003b) *Holbourne Island NP Management Plan*, QEPA, Brisbane.
QEPA (2003c) *Snapper Island NP and Marine Park: Visitor Information*, QEPA, Brisbane.
QGITB (Queensland Government Intelligence and Tourist Bureau) (1923) *The Great Barrier Reef of Australia: A Popular Account of its General Nature, Compiled by the GBRC*, QGITB, Brisbane.
QGTB (1931) *The Great Barrier Coral Reef, Queensland*, QGTB, Brisbane.
QGTB (1932) *Heron Island, Capricorn Group, Great Barrier Reef, Queensland*, QGTB, Brisbane.
QGTB (1940) *Winter Tours on the Great Barrier Reef*, Government Printer, Brisbane.
QNPWS (1996) *Keppel Bay Islands NP: Visitor Information*, QNPWS, Brisbane.
QNPWS (1998a) *Holbourne Island National Park Management Plan*, QNPWS, Brisbane.
QNPWS (1998b) *Michaelmas and Upolu Cays National Park Management Plan*, QNPWS, Brisbane.

QNPWS (1999a) *Capricornia Cays National Park and Capricornia Cays National Park (Scientific) Management Plan*, QNPWS, Brisbane.
QNPWS (1999b) *Hinchinbrook Island National Park Management Plan*, QNPWS, Brisbane.
QNPWS (1999c) *Lady Musgrave Island: Visitor Information*, QNPWS, Brisbane.
QPWS (2000) *Family Islands National Park Management Plan*, QPWS, Brisbane.
Queensland Marine Industries Ltd (1932), *Sharks and Turtles for Profit*, Queensland Marine Industries Ltd, Brisbane.
Ratcliffe, F. (1938) *Flying Fox and Drifting Sand: The Adventures of a Biologist in Australia*, Chatto and Windus, London.
Reclus, E. (1882) *Australasia*, ed. A. H. Keane, Virtue and Company, London.
Rees, H. (1962) *Australasia: Australia, New Zealand, and the Pacific Islands*, Macdonald and Evans, London.
Reid, F. (1933) 'Angling thrills in coral waters', *Cummins and Campbell's Monthly Magazine*, vol. 5, no 75, pp37–41.
Resource Assessment Commission (1993) *Coastal Zone Inquiry: Queensland Case Study: Coastal Zone Management in the Cairns Area*, Resource Assessment Commission, Canberra.
Reynolds, H. (1982) *The Other Side of the Frontier: Aboriginal Resistance to the European Invasion of Australia*, Penguin, Ringwood, Victoria.
Reynolds, H. (1987) *Frontier: Aborigines, Settlers and Land*, Allen and Unwin, Sydney.
Reynolds, H. (2003) *North of Capricorn: The Untold Story of Australia's North*, Allen and Unwin, Crows Nest, New South Wales.
Richards, H. C. (1937) 'Some problems of the Great Barrier Reef', *Journal and Proceedings of the Royal Society of New South Wales*, vol. 71, pp67–85.
Robertson, B. M. (2000) *Oral History Handbook*, 4th edition, Oral History Association of Australia (South Australia Branch) Inc., Adelaide.
Rohde, R. T. (1951) *A Visit to 'Boiling Springs': The Hope Vale (U.E.L.C.A.) Lutheran Mission (Re-Established), Cooktown, Queensland*, Lutheran Church of Australia, North Ipswich, Queensland.
Roughley, T. C. (1936) *The Wonders of the Great Barrier Reef*, Angus and Robertson, Sydney.
Rowe, G. (1865) *The Colonial Empire of Great Britain, Considered Chiefly with Reference to its Physical Geography and Industrial Productions: The Australian Group*, SPCK, London.
Royal Commission into Exploratory and Production Drilling for Petroleum in the Area of the Great Barrier Reef (1974) *Report of the Royal Commission*, vol. 2, AGPS, Canberra.
Saint-Smith, E. C. (1919) 'Geological survey report: rock phosphate deposit on Holbourne Island, near Bowen', *Queensland Government Mining Journal*, vol. 20, 15 March 1919, pp122–124.
Saville-Kent, W. (1890a) 'Bêche-de-mer and pearl-shell fisheries of northern Queensland', *QVP*, vol. 3, part 2, pp727–734.
Saville-Kent, W. (1890b) 'Fisheries, Wide Bay district', *QVP*, vol. 3, part 2, pp713–714.
Saville-Kent, W. (1890c) 'Pearl and pearl-shell fisheries of northern Queensland', *QVP*, vol. 3, part 2, pp703–712.
Saville-Kent, W. (1893) *The Great Barrier Reef of Australia: Its Products and Potentialities*, W. H. Allen, London.
Scriven, E. (1915) Annual Report, QBSES, 1914–15, *QPP*, vol. 2, p1175.
Scriven, E. (1916) Annual Report, QBSES, 1915–16, *QPP*, vol. 2, p1237.
Scriven, E. (1922) Annual Report, QBSES, 1920–21, *QPP*, vol. 2, p1034.

Semon, R. (1899) *In the Australian Bush and on the Coast of the Coral Sea: Being the Experiences and Observations of a Naturalist in Australia, New Guinea and the Moluccas*, Macmillan and Co., London.

Senior, W. (1877) *By Stream and Sea: A Book for Wanderers and Anglers*, Chatto and Windus, London.

Serventy, V. (1955) *Australia's Great Barrier Reef: A Handbook on the Corals, Shells, Crabs, Larger Animals and Birds, With Some Remarks on the Reef's Place in History*, Georgian House, Melbourne.

Sharp, J. P. (2000) 'Towards a critical analysis of fictive geographies', *Area*, vol. 32, no 3, pp327–334.

Skewes, T., Dennis, D., Koutsoukos, A., Haywood, M., Wassenberg, T. and Austin, M. (2004) 'Stock survey and sustainable harvest strategies for Torres Strait *beche-de-mer*', Final Report, CSIRO Marine Research and Australian Fisheries Management Authority, Cleveland, Queensland.

Smart, P. (1951) 'The dugong', *Walkabout*, 1 November 1951, pp34–35.

Smith, A. J. (1987a) 'An ethnobiological study of the usage of marine resources by two Aboriginal communities on the east coast of Cape York Peninsula, Australia', Unpublished PhD Thesis, Department of Zoology, JCU, Townsville.

Smith, A. J. (1987b) 'Usage of Marine Resources by Aboriginal Communities on the East Coast of Cape York Peninsula: Report to the GBRMPA, June 1987', Research Publication no 10, GBRMPA, Townsville.

Smith, J. R. (1916) Annual Report, QDHM, 1915, QPP, vol. 3, pp1635–1668.

Smyth, D. (1994) 'Understanding Country: The Importance of Land and Sea in Aboriginal and Torres Strait Islander Societies', Key Issue Paper no 1, Council for Aboriginal Reconciliation, AGPS, Canberra.

SPCK (c.1865) *Australia; A Popular Account of its Physical Features, Inhabitants, Natural History and Productions: With the History of its Colonization*, SPCK, London.

Spencer, G. L. and Meade, G. P. (1945) *Cane Sugar Handbook: A Manual for Cane Sugar Manufacturers and their Chemists*, 8th edition, John Wiley and Sons, New York.

Steers, J. A. (1938) 'Detailed notes on the islands surveyed and examined by the geographical expedition to the Great Barrier Reef in 1936', *Reports of the GBRC*, vol. 4, part 3, pp51–104.

Stevens, J. H. (1900) Annual Report, Inspector of Fisheries, 1899–1900, QVP, vol. 5, part 2, pp997–998.

Stevens, J. H. (1901) Annual Report, Inspector of Fisheries, 1900–1901, QPP, vol. 3, Part 2, pp1321–1325.

Stevens, J. H. (1902) Annual Report, Inspector of Fisheries, 1901–1902, QPP, vol. 3, pp965–968.

Stevens, J. H. (1905) Annual Report, Inspector of Fisheries, 1904–1905, QPP, vol. 2, pp1037–1042.

Stevens, J. H. (1906) Annual Report, Inspector of Fisheries, 1905–1906, QPP, vol. 2, pp1403–1424.

Stevens, J. H. (1908) Annual Report, Inspector of Fisheries, 1907–1908, QPP, vol. 3, pp904–907.

Stevens, J. H. (1909) Annual Report, Inspector of Fisheries, 1908–1909, QPP, vol. 2, pp927–932.

Stevens, J. H. (1910) Annual Report, Inspector of Fisheries, 1909–1910, QPP, vol. 3, pp924–928.

Stevens, J. H. (1911) Annual Report, Inspector of Fisheries, 1910–1911, *QPP*, vol. 3, pp1191–1193.
Stevens, J. H. (1913) Annual Report, Inspector of Fisheries, 1912–1913, *QPP*, vol. 3, pp1031–1033.
Stevens, J. H. (1914) Annual Report, Inspector of Fisheries, 1913–1914, *QPP*, vol. 3, pp974–975.
Stevens, J. H. (1918) Annual Report, Inspector of Fisheries, 1917–1918, *QPP*, vol. 1, pp1663–1665.
Stevens, J. H. (1919) Annual Report, Inspector of Fisheries, 1918–1919, *QPP*, vol. 3, pp340–344.
Stevens, J. H. (1920) Annual Report, Inspector of Fisheries, 1919–1920, *QPP*, vol. 2, pp569–572.
Stoddart, D. R., McLean, R. F., Scoffin, T. P. and Gibbs, P. E. (1978) 'Forty-five years of change on low wooded islands, Great Barrier Reef', *Philosophical Transactions of the Royal Society of London, Part B*, vol. 284, pp63–80.
Stoddart, E. J. (1933) 'The economic possibilities of the Great Barrier Reef', in Tilghman, D. C. (ed), *The Queen State: A Handbook of Queensland Compiled Under Authority of the Government of the State*, John Mills Himself, Brisbane, pp217–222.
Suggate, L. S. (1940) *Australia and New Zealand with Pacific Islands and Antarctica*, revised edition, George G. Harrap and Co., London.
Sunter, G. H. (1936) 'The dugong', *Walkabout*, 1 March 1936, pp47–48.
Sunter, G. H. (1937) *Adventures of a Trepang Fisher: A Record Without Romance, Being a True Account of Trepang Fishing on the Coast of Northern Australia, and Adventures Met in the Course of the Same*, Hurst and Blackett, London.
Taylor, G. (1925) *Australia in its Physiographic and Economic Aspects*, 4th edition, Clarendon Press, Oxford.
Tennant, D. (c.1940) *Marvels of the Great Barrier Reef, North Australia and New Guinea*, 2nd edition, Thomas Tennant, Chatswood, New South Wales.
Thomson, D. F. (1956) 'The fishermen and dugong hunters of Princess Charlotte Bay', *Walkabout*, vol. 22, no 11, pp33–36.
Thomson, D. F. (1985) *Donald Thomson's Mammals and Fishes of Northern Australia*, ed. J. M. Dixon and L. Hurley, Nelson, Melbourne.
Thomson, J. (1966) *The Great Barrier Reef*, Nelson Doubleday, Crow's Nest, New South Wales.
Thomson, J. (ed.) (1989) *Reaching back: Queensland Aboriginal People Recall Early Days at Yarrabah*, Aboriginal Studies Press, Canberra.
Thorne, E. (1876) *The Queen of Colonies; Or, Queensland As I Knew It*, Samson Law, Marston, Searle and Rivington, London.
Tilghman, D. C. (1933) *The Queen State: A Handbook of Queensland Compiled Under Authority of the Government of the State*, John Mills Himself, Brisbane.
Travers, M. (2001) *Qualitative Research Through Case Studies*, Sage Publications, London.
Uthicke, S. (2004) 'Overfishing of holothurians: lessons from the Great Barrier Reef', in Lovatelli, A., Conand, C., Purcell, S., Uthicke, S., Hamel, J.-F. and Mercier, A. (eds), *Advances in Sea Cucumber Aquaculture and Management*, FAO Fisheries Technical Paper 463, FAO, Rome, pp163–171.
Uthicke, S. and Benzies, J. A. H. (2000) 'Effect of bêche-de-mer fishing on densities and size structure of *Holothuria nobilis* (Echinodermata: Holothuroidea) populations on the Great Barrier Reef', *Coral Reefs*, vol. 19, pp271–276.

Uthicke, S., Welch, D. and Benzie, J. A. H. (2004) 'Slow growth and lack of recovery in overfished holothurians on the Great Barrier Reef: evidence from DNA fingerprints and repeated large-scale surveys', *Conservation Biology*, vol. 18, no 5, pp1395–1404.

Veron, J. E. N. (2009) *A Reef in Time: The Great Barrier Reef From Beginning to End*, Belknapp Press of Harvard University Press, Cambridge, MA.

Wachenfeld, D. R. (1995) 'Report of the Historical Photographs Project of the Great Barrier Reef Marine Park Authority', Unpublished Report, GBRMPA, Townsville.

Wachenfeld, D. R. (1997) 'Long-term trends in the status of coral reef-flat benthos: the use of historical photographs', in Wachenfeld, D. R., Oliver, J. K. and Davis, K. (eds), *State of the Great Barrier Reef World Heritage Area Workshop: Proceedings of a Technical Workshop Held in Townsville, Queensland, Australia, 27–29 November 1995*, Workshop Series No. 23, GBRMPA, Townsville, pp134–148.

Wachenfeld, D. R., Oliver J. K. and Morrissey J. I. (eds) (1997), *State of the Great Barrier Reef World Heritage Area 1998*, GBRMPA, Townsville.

Walsh, A. (1987) *Lady Elliot: First Island of the Great Barrier Reef*, Boolarong Publications, Brisbane.

Wandandian (R. Dyatt) (1912) *Travels in Australasia*, Birmingham, Cornish Brothers.

Welsby, T. (1905) *Schnappering and Fishing in the Brisbane River and Moreton Bay Waters*, The Outridge Printing Co., Brisbane.

Welsby, T. (1907) *Early Moreton Bay Recollections*, The Outridge Printing Co., Brisbane.

Welsby, T. (1931) *Sport and Pastime in Moreton Bay*, Simpson Halligan and Co., Brisbane.

Wilkinson, C. (ed.) (2000), *Status of Coral Reefs of the World: 2000*, AIMS, Townsville.

Williams, D. McB. (2001) 'Impacts of terrestrial run-off on the Great Barrier Reef World Heritage Area', Report to CRC Reef, CRC Reef Research Centre, Townsville.

Williams, D. McB., Roth, C. H., Reichelt, R., Ridd, P., Rayment, G. E., Larcombe, P., Brodie, J., Pearson, R., Wilkinson, C., Talbot, F., Furnas, M., Fabricius, K., McCook, L., Hughes, T., Hoegh-Guldberg, O. and Done, T. (2002) 'The current level of scientific understanding on impacts of terrestrial run-off on the Great Barrier Reef World Heritage Area', Consensus Statement, CRC Reef Research Centre, Townsville.

Wood, A. W., Bramley, R. G. V., Meyer, J. H. and Johnson, A. K. L. (1997) 'Opportunities for improving nutrient management in the Australian sugar industry', in Keating, B. A. and Wilson, J. R. (eds), *Intensive Sugarcane Production: Meeting the Challenges Beyond 2000*, CAB International, Wallingford, Oxfordshire, pp243–266.

Woodland, D. J. and Hooper, J. N. A. (1977) 'The effect of human trampling on coral reefs', *Biological Conservation*, vol. 11, no 1, pp1–4.

Yonge, C. M. (1930) *A Year on the Great Barrier Reef: The Story of Corals and of the Greatest of their Creations*, Putnam, London.

Index

A. and I. Products Pty Ltd 88
Aboriginal people 12, 45, 55, 58, 62, 87, 90, 91, 105, 110, 205; see also Indigenous Australians
access tracks and channels 13, 27, 29, 30, 184–8, 198, 225
acidification see ocean acidification
Acropora 39, 40
Agent-General for Queensland 23
agriculture 7, 46, 48, 53, 135, 156; see also sugar cane
AIMS (Australian Institute of Marine Science) 19
airstrips 201, 212–13
Alexandra Reef 37, 38, 167, 172, 173, 179–80
Amity Point 96, 98, 102–4
ANFB (Australian National Film Board) 29
Anglo-Australian Guano Company 158
Applied Ecology Pty Ltd 87
aquaculture 4, 53, 151
aquarium fish collecting 6, 125–6; see also fishing
Arlington Reef 146, 175
Astrolabe 22
Atherton Tableland 50, 206,
A.U.S.N. Company 60, 68, 161
Australian Airlines 196
Australian Institute of Marine Science see AIMS
Australian National Film Board see ANFB
Australian Turtle Company Limited 80
Australian Whaling Commission 114

Badu Island 90, 106
bananas 4, 213–14
Banfield, Edmund 22, 103, 109, 135, 212
Barnard Islands 56, 166, 168, 171–2, 180, 225
Barrier Reef Airways 196
Barrier Reef Islands Pty Ltd 197
Barrier Reef Trading Company 80–1
baselines: ecological 6, 8, 227; shifting 227
beacon 10, 155–6, 161, 165, 225, 227
Beanley Island 39
Beaver Reef 167, 174, 176
bêche-de-mer 6, 9, 10, 13, 43, 48, 51, 53, 55–71, 74–6, 103, 106, 117, 132, 134–5, 203, 224
Bennett, Isobel 10, 18, 26, 28, 41, 147, 185, 186
bioerosion 34, 36, 37
birds 26, 158, 175, 177–8, 190, 208, 216–18, 226
Blackwood Channel 155
Bloomfield River 59
Boigu Island 87, 107–8
bombing see target practice
books (historical) 10, 11, 15–17, 19, 21–3, 129, 156
Bowen 6, 39, 47, 60, 68, 109, 118, 121, 144, 160–1, 202, 205
Bramble Cay 77, 82–3, 87–9
Brammo Bay 213
Brampton Island 132, 133, 194, 202, 206, 210, 213–15, 218
Brisbane 11, 46–8, 78, 79, 81, 83, 100, 107, 118, 120, 122, 161, 164, 172, 177, 196
Brisbane Fish Agency Company 78
Brisbane Preserving Works 78
Brisbane River 196

Burdekin River 47
Burrum Heads 103, 105, 110
Burrum River 104

Caesalpinia bonduc 210–11
Cairns 6, 11, 26, 33, 48, 50, 52, 54, 56, 57, 68–9, 107–8, 119, 121–2, 124, 126, 137, 140–142, 147–8, 173–6, 180, 188, 194, 225
Cairns District Canegrowers Association 177
Cairns Harbour Board 39, 130, 192–3,
calcium 165
cane toad 50, 218
Cape Bedford 97, 98, 106; *see also* Hope Vale
Cape Direction 109; *see also* Lockhart River
Cape Melville 59
Cape Moreton 18, 30, 130, 150, 187
Cape Sidmouth 109
Cape Tribulation 37, 40, 167, 191,
Cape York 3, 54, 68, 91, 165
Cape York Peninsula 108
Capricorn-Bunker Group 11, 29, 33, 76–84, 92, 148, 158–60, 162, 194, 208
carbonate 5
Cardwell 48, 96–8, 144, 165
Carlisle Island 205, 212
Casuarina 190, 208, 211, 212
catchment 5, 43, 52; *see also* GBRCA
Cato's Bank 156
cattle grazing 4, 6, 217; *see also* pastoralism
cays *see* islands
cedar 48, 53
Central Queensland Meat Export Company Limited 81
channel 13, 26, 27, 29, 30, 34, 91, 1841–7, 198, 225
Chemist Roush 107
Cherbourg 106
Chillagoe 50, 171, 176
Chillagoe Railway and Mines Company 168
Cid Harbour 132, 133, 189
clams *see* giant clams
climate change 5, 228
closer settlement 12, 50, 53
coastal development 6, 13, 26, 43
coastal protection 5

Coconut Island 87, 89, 90
coconut palm 24, 27, 51, 201–6, 214, 215, 218, 219, 225
co-management 4
Commonwealth Government 11, 49, 65, 150
Commonwealth of Australia Productivity Commission 6
conservation 1, 4, 43, 52, 62, 87, 89, 90, 95, 101, 105, 134, 135, 150
conservation tillage 51
continental drift 44
continental shelf 12, 34–5, 41
Cook, James 21, 45, 143
Cooktown 39, 48, 56–8, 62, 68, 117, 146, 172, 202, 225
copra 48, 201, 205
coral: bleaching 5, 139; collecting 6, 9, 16, 27, 30, 33, 38, 54, 128–43, 151; coral sand 50, 133, 166, 171–2, 177; disease 5; micro-atolls 40, 185; mining 11, 26, 27, 30, 31, 38, 50, 52, 54, 129, 164–80, 225; transplantation 138
coralline algae 34
Coral Sea 45, 188
coral trout *see* fishing
cotton 48
Crooke, Ray 27
crown-of-thorns starfish 30
CSIRO (Commonwealth Scientific and Industrial Research Organisation) 115, 116
Cumberland Islands 132, 136
curios 87, 88, 90, 118, 129–31, 134, 149, 151
cyclones *see* tropical cyclones

Daintree 167, 170
Darnley Island 87, 88, 165
Day Dream Island 196
deforestation 12, 48, 213, 215
Denman Island 136, 217
Dingo Beach 6
diving 55, 57, 62, 64, 68, 69
Doomadgee 106
Double Island, 142–3, 194
drainage 12, 51, 213
drainage basins 4
dredging 6, 24, 25, 185, 224
dugongs 4, 10, 95–6, 110–111, 224: dugong oil 95–6, 98–108, 110;

fishing 51, 95–108; Indigenous hunting, 83, 87, 88, 108–110
Dunk Island 22, 52, 56, 133, 176, 194–8, 212–13, 215, 216, 218
Dunk Island Tropical Tourist Board 216

ecological threshold *see* threshold
economic development 4, 46, 48, 134
El Arish 50
Ellison Reef 52, 164, 177
Endeavour 21, 45, 73, 77, 120, 129
Endeavour Reef 59
Endeavour River 77
Endeavour Strait 62–3
environmental management 14, 31
environmental threshold *see* threshold
erosion 6, 12, 34, 51–2, 93, 159, 166, 178, 185, 206, 210, 212, 215; *see also* bioerosion
eutrophication 5
expeditions; *see also* Great Barrier Reef Expedition

Fairfax Islands 14, 137, 146–7, 151, 159, 184, 189–91, 195, 198, 208–10, 216, 218, 225–7
fauna *see* islands
feral pigs *see* islands
fertiliser 49–50, 52, 80, 118, 156, 160, 166–7, 171–2, 174, 177
fiction (historical) 16, 21–3
film (historical) 11, 16, 18, 19, 27, 29, 86, 125, 139, 217
fire 45, 201, 203, 212, 216, 226
First World War, 50, 60, 66, 67, 76, 103, 160, 168–9
fish 6, 30, 113, 118–27, 132: aquarium 6, 125–6; live 126; *see also* fishing
fishing 119–26: commercial 4, 228; dynamite 120–1, 225; recreational 118, 121, 194, 196; spearfishing 124–5; trawling 122
Fitzroy Island 56, 121, 165–6
Fitzroy River 31, 47, 78, 81–2
Flinders, Matthew 2, 21, 47
Ford Sherrington Ltd 118
Four Mile Beach 6
Frankland Islands 56, 137, 172

GBRC (Great Barrier Reef Committee) 22, 23, 39, 83, 103, 134–6, 144, 158, 185, 194

GBRCA (Great Barrier Reef Catchment Area) 3, 4, 49, 51–3
GBRMP (Great Barrier Reef Marine Park) 1–3, 11, 52, 164, 226
GBRMPA (Great Barrier Reef Marine Park Authority) 1, 16, 27, 30
GBRWHA (Great Barrier Reef World Heritage Area) 1, 3–5, 10, 52–3
Genami Gia Turtle Fishing 82
geomorphology 12, 34–8, 41, 155–6, 162, 225
giant clams 140, 143, 147–51
Gladstone 46, 52, 54, 81–3, 122, 140, 165, 194, 196
glass-bottomed boats 124, 195–7
global warming *see* climate change
goats 201, 206–11, 215, 217–1-9, 225
gold mining 46, 48
Goold Island 37, 38, 215
Goondi 50, 166
government records 11, 17, 19, 21, 25–6, 55
government reports 11, 16, 17, 21, 23–4, 55
Grassy Island 207, 217
Great Barrier Reef Catchment Area *see* GBRCA
Great Barrier Reef Committee *see* GBRC
Great Barrier Reef Expedition 10, 22, 39
Great Barrier Reef Fisheries Ltd 81, 117, 172
Great Barrier Reef lagoon 5–6, 35, 155
Great Barrier Reef Marine Park *see* GBRMP
Great Barrier Reef Marine Park Authority *see* GBRMPA
Great Barrier Reef Protection Interdepartmental Committee Science Panel 6
Great Barrier Reef Province 2
Great Barrier Reef Region 2
Great Barrier Reef World Heritage Area *see* GBRWHA
Green Island 14, 29, 39, 51, 52, 56, 60–1, 121, 123–4, 128, 130–1, 137–8, 142, 144–51, 171, 175, 184, 189, 191–8, 212, 213, 218

guano 13, 26, 29, 51, 53–4, 117, 155–62, 201, 206, 207, 211, 218, 225, 227–8
guavas 202, 213

habitat degradation 6, 13, 33, 51, 53, 72, 92, 95, 113, 120, 127, 192, 211, 226
Halifax Bay 37–8
Haslewood Island 214, 217
Hayles Magnetic Pty Ltd 123, 192–4
Hayman Island 51, 122, 128, 131–2, 145, 150, 189, 192, 197–8
Hedley, Charles 134
Henning Island 212, 214
Herbert River 47
Heron Island 6, 14, 26, 28, 41, 51, 77, 80–1, 84–7, 124, 128, 131–2, 137–9, 142, 145–6, 148, 150–1, 184–6, 191–2, 194–8, 216–18
Hervey Bay 4, 10, 96–100, 103–5, 110
High Island 137, 172
Hinchinbrook Channel 103,
Hinchinbrook Island 6, 59, 214–16, 218
Historical Photographs Project 18, 27–8
Hobbs, William 96, 98, 100
Holbourne Island 14, 37, 39–40, 155, 160–2, 215, 225–6
Holocene 1, 12, 34–5, 38, 41, 44
Hope Vale 106–7, 242
Hoskyn Islands 137, 160, 195, 210
humpback whales 10, 13, 113–17, 126, 224–5, 227
hutchinson Island 171–2, 174–5

IIB (Island Industries Board) 106
Indigenous Australians 1, 4, 12, 13, 44–5, 53, 58, 72, 87, 90–2, 95, 102–3, 105–6, 108–10, 203
infrastructure 13, 24, 27, 117, 159, 165, 184, 192, 197–8, 201, 212, 225
Ingham 50, 56, 57
Ingram Island 39
Inner Passage 47
Innisfail 48, 50, 52, 137, 164, 168, 171–80
insecticide 50–1
interviewing see qualitative interviewing
Isis River 103

Island Industries Board see IIB
island: fauna 216–18; management plans 20, 26; vegetation 13, 29, 201, 206–15, 225–6
IUCN 72, 73, 92

J. T. Arundel and Company 158
Jessie Island 171, 175
jetties 39, 124, 158, 159, 178, 192–3, 196–7
Jukes, Joseph Beete 47, 77, 130, 144, 155–6, 158

Kangaroo Reef 37
Kauri pine 202
Keppel Bay 78
Kings Reef 38, 164, 168, 175–6, 178–81, 225
Kubin Island 88, 90
Kurrimine Beach 6, 179

Lady Elliot Island 14, 39, 155, 157–60, 162, 207, 213, 225–6
Lady Musgrave Island 6, 14, 84, 86, 132, 136–7, 145–8, 150–1, 184–5, 187, 189 194–5, 197–8, 207–11, 215–16, 218, 225–6
land clearance 5, 50, 53, 135, 213
land use 4, 5, 43, 53
Langford Reef 6, 144
Lantana 54, 201, 214–15, 219, 226
leaflets (historical) 11, 16, 17, 19, 21, 23
lime 50, 129, 156, 161, 164–77, 180, 225
Lindeman Island 91, 121, 132, 135, 196, 197, 209, 210, 212–14, 217
Linnaeus 129
Lizard Island 52, 56, 70, 124–6, 147–8, 151, 197, 213, 218
Lockhart River 106, 108, 109
Long Island 88, 136, 207, 215, 217
Low Isles 10, 22, 28, 29, 39, 130, 148, 175
Lucinda 174, 176

Mabuiag Island 88, 108–9
Mackay 6, 48, 50, 57, 69, 70, 102, 135, 140, 173–4, 194, 202, 205, 215, 225
Mackay Cay 39
Mackay Reef 189
Magnetic Island 14, 27, 52, 123, 192–4, 198, 212, 214
mangoes 202, 214

mangroves 39, 51, 58, 69, 87, 109, 172
maps (historical) 15, 16, 18, 19, 21, 27, 29, 30, 31, 45, 141, 174
Maria Creek 171
marine turtles *see* turtles
marlin *see* fishing
Maryborough 48, 100, 120
Master Foods Corporation 87
Masthead Island 78, 80, 84–5, 130–2, 150, 172, 214
Mather, Patricia 185
MCMC (Mossman Central Mill Company) 167–8, 172
Michaelmas Cay 26, 39, 147, 151, 156, 160, 162, 175, 188, 217
Michaelmas Reef 146, 148
Middle Island 37
military target practice *see* target practice
mines (military) 188–9
Molle Islands 132, 144, 192, 193, 215, 217: Mid Molle 136, 217; North Molle 207, 217; South Molle 84, 136, 214, 217, 218; West Molle 136
monitoring (scientific) 6–8, 52, 61, 73, 90, 92, 111, 116–17, 126, 128, 141, 180, 225–8
Monkman, Noel 29, 86, 135, 137–9
Moreton Bay 4, 10, 43, 46, 53, 55, 62, 77, 78, 95–105, 110–11,
Moreton Island 114
Mornington Island 87–8, 108–9
Mossman 50, 164, 166, 172, 177, 179
Mossman Central Mill Company *see* MCMC
Mossman River 168
mother-of-pearl *see* pearl-shell
Mourilyan 50, 166
Mowbray River 166, 172
Murray Island 87–8, 201

NADC (Northern Australia Development Committee) 66
narrative approach 7–8
Nathan, Matthew 134, 173
National Parks Association of Queensland 135, 146
naturalists 9, 16, 47, 129, 130, 132, 177, 191; *see also* NQNC
navigation 10, 12, 29, 44; *see also* Raine Island
newspapers (historical) 15–17, 20–1, 26

Night Island 39
nitrogen 5, 49, 50
noise pollution 113, 126
North Keppel Island 206, 207, 210, 215, 217, 218
North Palm Island 207
North Queensland Naturalists' Club *see* NQNC
North Reef 29, 184, 187, 188, 198
North Stradbroke Island 96, 101
North West Island 6, 14, 51, 80–1, 84, 132, 145, 146, 155, 159–60, 162, 217, 225, 226
Northern Australia Development Committee *see* NADC
Northumberland Islands 132
NQNC (North Queensland Naturalists' Club) 120, 135, 217
nutrients 4–6, 12, 37, 50–2, 166, 212

ocean acidification 5, 228
oil drilling 52
One Tree Island 139–40
Opuntia see prickly pear
Orman's Reef 108
Orpheus Island 40, 148–50, 207
Otter Reef 142
Oyster Cay 121, 160, 174–5, 177–8, 225
oysters 132, 196; *see also* pearl-shell

Palm Island 37, 82, 83, 91, 98, 106, 202, 205–7
Pandora Cay 77
Pastoral Supplies Ltd 107
pastoralism 43, 46, 48, 217
pearl *see* pearl-shell
pearl-shell 10, 13, 43, 51, 55, 57–8, 61–70, 74–6, 79, 103, 107, 134, 224
Pelican Banks 96, 102, 104
Percy Isles 132, 206, 207, 210
pesticide 50, 51, 215
phase shift (ecological) 36, 37
phosphate *see* rock phosphate
phosphorus 5, 49, 50
photographs (historical) 8, 10, 15, 16, 18, 19, 20, 26–8, 31, 129, 185, 210
Picnic Bay 192
pigs *see* islands
Pisonia 80, 159, 190, 208–9, 211, 225
plantations *see* islands

Pleistocene 38, 44
pollution 5, 6, 52, 92, 113, 126
Port Albany 165
Port Curtis *see* Gladstone
Port Denison *see* Bowen
port development 224
Port Douglas 6, 40, 52, 172, 173, 179
Port Hinchinbrook 52
prickly pear 201, 214–15
Prince of Wales Island 65
Princess Charlotte Bay 65, 91, 109, 189
Proserpine 194

Qantas 140
QBSES (Queensland Bureau of Sugar Experimental Stations) 49–50, 165–6
QDAIA (Queensland Department of Aboriginal and Island Affairs) 17
QDHM (Queensland Department of Harbours and Marine) 11, 17, 23, 24, 76, 82, 86, 116, 118, 119, 124, 129, 131, 142, 145, 164, 170–1, 175, 178, 180, 185
QDNA (Queensland Department of Native Affairs) 7, 23, 24, 87
QEPA (Queensland Environmental Protection Agency) 17, 19, 156, 164, 201
QGITB (Queensland Government Intelligence and Tourist Bureau) 23
QGTB (Queensland Government Tourist Bureau) 23, 30, 83, 86, 135–6, 145, 191, 195, 196
QNPWS (Queensland National Parks and Wildlife Service) 83, 159, 209, 210, 217
QPWS (Queensland Parks and Wildlife Service) 20, 26
QSA (Queensland State Archives) 16, 19, 23–6, 28, 30, 164, 172, 176, 180, 201
QSPCA (Queensland Society for the Prevention of Cruelty to Animals) 86, 216
qualitative interviewing 7, 30, 31, 65
Queensland Airlines 184
Queensland Bureau of Sugar Experimental Stations *see* QBSES
Queensland Conservation Council 192
Queensland Department of Aboriginal and Island Affairs *see* QDAIA
Queensland Department of Agriculture and Stock *see* QDAS
Queensland Department of Harbours and Marine *see* QDHM
Queensland Department of Native Affairs *see* QDNA
Queensland description and travel literature 17, 21, 22
Queensland Environmental Protection Agency *see* QEPA
Queensland Fish Board 82
Queensland Government Intelligence and Tourist Bureau *see* QGITB
Queensland Government Tourist Bureau *see* QGTB
Queensland Marine Industries Ltd 81, 118
Queensland National Parks and Wildlife Service *see* QNPWS
Queensland Parks and Wildlife Service *see* QPWS
Queensland Royal Mail Line 47
Queensland Society for the Prevention of Cruelty to Animals *see* QSPCA
Queensland State Archives *see* QSA

Rabbit Island 205
rabbits 218
Raine Island 10, 29, 92, 149, 155–9, 161–2, 165, 225, 227
rainforest 50, 53, 167, 214
Rattlesnake 47, 56
rays *see* stingrays
reclamation (land) 52, 213
Redbill Island 144
reef-walking 13, 129, 137, 184, 191–2, 198, 225
Repulse Bay 98, 102,
Repulse Islands 213–14, 217
resilience (ecosystem) 5, 228
ringbarking 212
riparian vegetation 51
rock phosphate 13, 51, 155–6, 160–2, 201, 225
Rockhampton 46, 47, 77, 79, 81
Rodds Bay 96–9
Royal Zoological Society of New South Wales 84
runoff 4–6, 12, 50–2, 110, 212

Saddleback Island 39, 207
St Bees Island 132, 217

St Helena Island 96
Sandpiper Reef 167, 174
Saville-Kent, William 22, 27, 39, 55–8, 63–4, 70, 73–4, 78, 91, 101–2, 108, 117, 130
scientific monitoring *see* monitoring
sea country 4
sea cucumbers *see bêche-de-mer*
sea surface temperature 34
seabirds *see* birds
Seaforth Island 136, 202
seagrass 5–6, 95–6, 100, 103, 110, 197
sea-level 5
Second World War 11, 27, 66, 69, 76, 106–7, 114, 121, 176–7, 180, 188–90
sediment 4–6, 12, 34–8, 40–1, 51–2, 185, 212
sharks 13, 113, 117–19, 122, 126–7
Shaw Island 133, 203
shell collecting 6, 9, 27, 33–4, 54, 128–9, 143–8, 150–1, 191, 227
shifting baselines *see* baselines
shipping 4, 6, 47, 52, 135, 161, 188, 189, 224
Sir Charles Hardy Island 202, 204
Snapper Island 27, 30, 164–5, 167–71, 179–80, 213, 225
soil erosion 12, 51–2, 212
spearfishing *see* fishing
Starcke River 106
Stephen Island 84
Stephens Island 214
Stewart River 109
stingrays 122
Stone Island 37, 39–41, 132, 144, 202
storms *see* tropical cyclones
South Australian Company 114
South Johnstone 50, 166
Sudbury Cay 174
Sudbury Reef 68
sugar cane 4, 12–13, 43, 48–53, 160, 164, 165, 171–2, 176, 180
Swain Reefs 68, 69
swamps 12, 51, 52, 172, 213

Tangalooma 114–15, 117, 126, 227
Tanner and Kenny Contractors 174–5, 177–8
target practice 13, 33, 34, 184, 188–9, 191, 198, 201, 208, 225, 227
Temple Bay 109
terrestrial runoff *see* runoff

Thalassia 39
threshold (ecological) 5, 35–7, 41, 181
Thursday Island 57, 58, 62, 63, 106–8
timber-getting 43, 48, 212
Tin Can Bay 96, 98, 100
tobacco 52
Toogoom 103, 110
Torres Strait Islanders 12, 55, 58, 62, 87, 90–2, 105, 110; *see also* Indigenous Australians
tortoise-shell 13, 24, 72–6, 80, 87, 92
tourist resorts 72, 81, 84–6, 92, 117, 121–4, 130–2, 135–8, 144–6, 148, 151, 191–8, 201, 212, 217–18, 225
Townsville 6, 11, 16, 33, 38, 46, 54, 56, 57, 60, 121–2, 140, 142, 161, 176, 194, 225
trawling *see* fishing
trepang see bêche-de-mer
Trinity Bay 109
trochus 13, 55, 62, 68–70, 103, 106, 135, 212, 224
tropical cyclones 6 12, 33, 34, 37–41, 144, 192
tropical fruit 4, 52, 214
Tryon Island 6, 160
Tully 50, 194
Tully River 91
turtles: farming 13, 86–90; fisheries 13, 27, 31, 51, 54, 72, 76–86, 92, 106, 224; flatback 72–3, 85; green 24, 72–3, 76–89, 92, 162, 224, 227; hawksbill 72–6, 87–92, 224; Indigenous hunting 90–2; leatherback 72–3, 91; loggerhead 72–3; olive ridley 72–3; turtle-riding 84–6; turtle soup 51, 77–8, 80–1

UNESCO (United Nations Educational, Scientific and Cultural Organization) 1
Upolu Cay 156, 160, 162, 171, 174–5, 177–8, 184, 189, 198, 225
urban development 4
urbanisation 53

vegetation *see* islands
von Linné, Carl *see* Linnaeus

Wallaby Reef 37
Warraber Island 88

water quality 5–6, 33, 38, 51, 89
Western Islands 89
Wet Tropics World Heritage Area *see* WTWHA
Whale Products Pty Ltd 114, 116–17
whales *see* humpback whales
whaling 27, 97, 113–17, 126, 225
Wistari Reef 138–9, 142, 146, 148, 150–1
Whitlam, Gough 89
Whitsunday Islands 6, 11, 52, 54, 84, 118, 122, 132, 136, 144–5, 148, 176, 194, 196, 202, 204, 207, 212, 214–15, 217
Whitsunday Passage 132, 214

Wide Bay 96–102
Wildlife Preservation Society of Queensland 148, 216
Woorabinda 106
World Heritage Area *see* GBRWHA
Wreck Reef 56, 156
Wright, Judith 26
WTWHA (Wet Tropics World Heritage Area) 52
Wynnum 98, 104, 118

Yam Island 87, 90
Yarrabah 106, 165, 194
Yorke Island 87–8
Yule Point 28, 151, 166, 172